Name: ____________________

11+

Non-Verbal Reasoning

Study and Revision Guide

Sarah Collins & Peter Francis

The Publishers would like to thank the following for permission to reproduce copyright material.

Photo credits

All photos are the authors' own.
Thanks to Edward Moon for permission to use his picture on page 47.

Orders: please contact Bookpoint Ltd, 130 Milton Park, Abingdon, Oxon OX14 4SB. Telephone: (44) 01235 827720. Fax: (44) 01235 400454. Email education@bookpoint.co.uk Lines are open from 9 a.m. to 5 p.m., Monday to Saturday, with a 24-hour message answering service. Visit our website at www.galorepark.co.uk for details of other revision guides for Common Entrance, examination papers and Galore Park publications.

ISBN: 978 1 4718 4925 1

First published in 2016 by

Galore Park Publishing Ltd,

An Hachette UK Company

Carmelite House

50 Victoria Embankment

London EC4Y 0DZ

www.galorepark.co.uk

Impression number 10 9 8 7 6 5 4 3 2

Year 2020 2019 2018

Typeset in India

Illustrations by Peter Francis

Printed in India

A catalogue record for this title is available from the British Library.

Contents and progress record

Use this page to plot your revision. Colour in the boxes when you feel confident with the skill and note your score and time for each test in the boxes.

How to use this book

Introduction

This book has been written to help you discover all about Non-Verbal Reasoning. It may be a subject you have never heard of, but that's nothing to worry about. The questions all involve skills you will have come across in Maths, just in a different way.

You will learn about:

- different kinds of questions you may see
- simple ways to solve the problems
- how to tackle difficult questions
- experts' tips on timing.

Use the book in the best way to help you learn. Work through the pages with a friend or parent or on your own, then try the questions and talk about them afterwards. Each skill or type of question takes up two pages so you can learn something new in half an hour. You are more likely to remember the skills and enjoy learning them in short bursts rather than spending an entire afternoon when you are tired, so try setting time aside after school two or three days a week. You may be surprised how quickly you learn.

Pre-Tests and the 11+ entrance exams

The Galore Park 11+ series is designed for Pre-Tests and 11+ entrance exams for admission into independent schools. These exams are often the same as those set by local grammar schools, too. There are now several different kinds of Non-Verbal Reasoning tests and it is likely that if you are applying to more than one school, you will encounter more than one type of test. These include:

- Pre-Tests delivered on-screen
- 11+ entrance exams in different formats from GL (Granada Learning) and CEM (Centre for Evaluation and Monitoring)
- 11+ entrance exams created specifically for particular independent schools.

Tests are designed to vary from year to year. This means it is very difficult to predict the questions and structure that will come up, making the tests harder to revise for. However, it's worth remembering that a great many of the questions require similar skills, and it is these skills that are introduced and practised throughout this *Study and Revision Guide*.

To give you the best chance of success in these assessments, Galore Park has worked with 11+ tutors, independent school teachers, test writers and specialist authors to create this *Study and Revision Guide*. The book covers the main types of questions now typically occurring and offers practice to increase your speed at answering the questions required in the most challenging examinations out there.

The learning ladders

These ladders appear throughout the book to show how skills in Non-Verbal Reasoning link to the familiar areas of Maths you have learned about in school. Your ability to problem-solve will develop as you step up the ladder, bringing together all your knowledge to solve the most challenging questions by the time you reach the top.

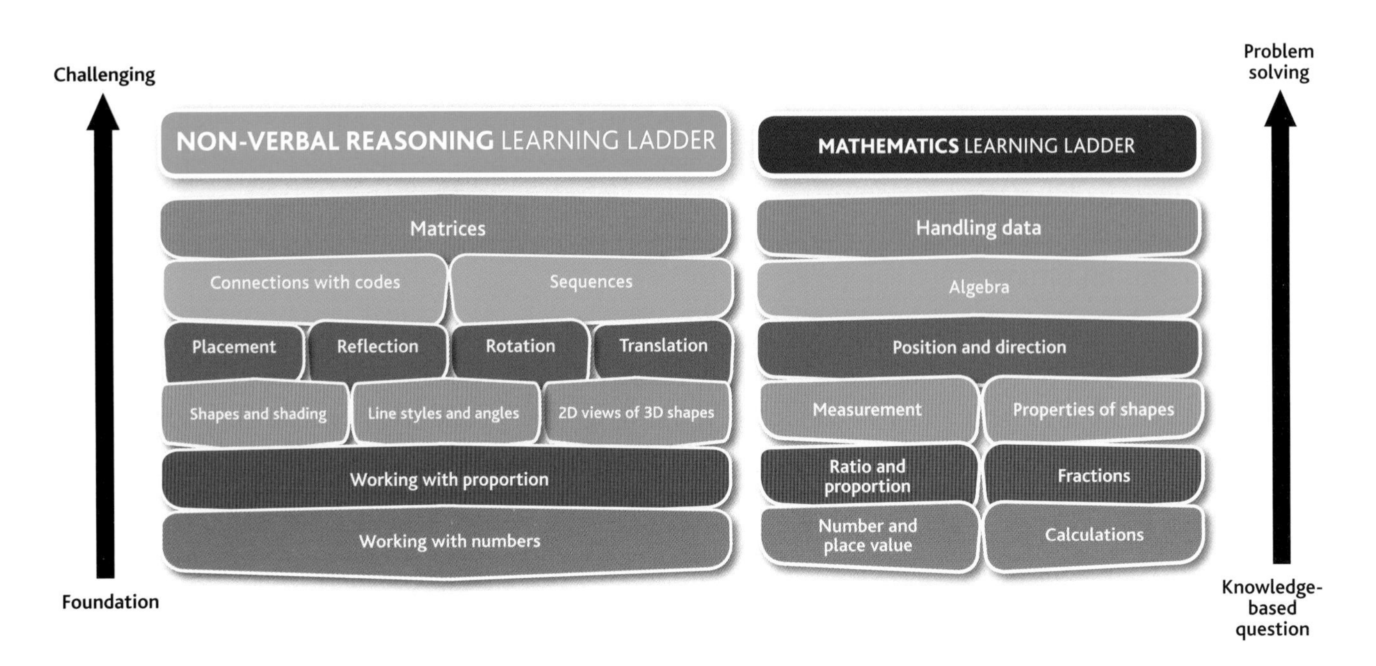

Working through the book

The **contents and progress record** helps you to keep track of your progress. When you have finished one of the learning spreads or tests, turn back to these introductory pages and complete them by:

- colouring in the 'Revised' box on the record when you are confident you have mastered the skill
- adding in your test scores and time to keep track of how you are getting on and work out which areas you may need more practice in.

The book is divided into three **parts**. *Part 1* outlines the basic skills in the learning ladders and introduces some of the questions you may encounter. *Part 2* tells you more about the individual skills. *Part 3* brings together these skills in some of the more challenging question types.

Chapters link together types of questions testing similar skills.

Chapter introductions explain how the skills link in with work you have done in Maths and provide revision tips to boost your Non-Verbal Reasoning skills.

Thinking skills and games are provided for you to practise with your friends and family to build your skills and to test everybody else's at the same time!

Learning spreads in Chapters 2 to 6 introduce one type of skill or question across two or four pages.

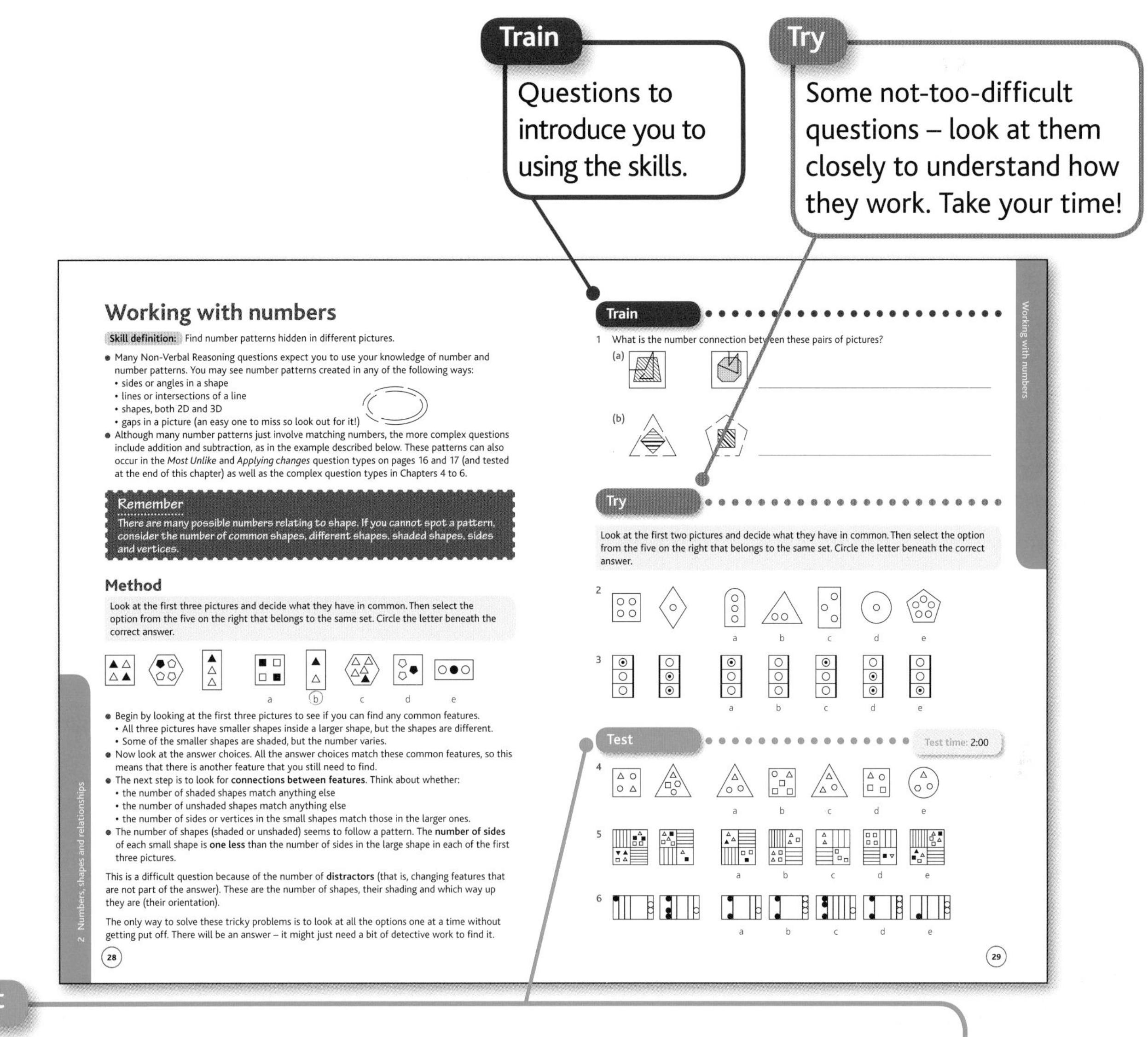

Working with numbers

Skill definition: Find number patterns hidden in different pictures.

- Many Non-Verbal Reasoning questions expect you to use your knowledge of number and number patterns. You may see number patterns created in any of the following ways:
  - sides or angles in a shape
  - lines or intersections of a line
  - shapes, both 2D and 3D
  - gaps in a picture (an easy one to miss so look out for it!)
- Although many number patterns just involve matching numbers, the more complex questions include addition and subtraction, as in the example described below. These patterns can also occur in the *Most Unlike* and *Applying changes* question types on pages 16 and 17 (and tested at the end of this chapter) as well as the complex question types in Chapters 4 to 6.

Remember
There are many possible numbers relating to shape. If you cannot spot a pattern, consider the number of common shapes, different shapes, shaded shapes, sides and vertices.

Method

Look at the first three pictures and decide what they have in common. Then select the option from the five on the right that belongs to the same set. Circle the letter beneath the correct answer.

a b c d e

- Begin by looking at the first three pictures to see if you can find any common features.
  - All three pictures have smaller shapes inside a larger shape, but the shapes are different.
  - Some of the smaller shapes are shaded, but the number varies.
- Now look at the answer choices. All the answer choices match these common features, so this means that there is another feature that you still need to find.
- The next step is to look for **connections between features**. Think about whether:
  - the number of shaded shapes match anything else
  - the number of unshaded shapes match anything else
  - the number of sides or vertices in the small shapes match those in the larger ones.
- The number of shapes (shaded or unshaded) seems to follow a pattern. The **number of sides** of each small shape is **one less** than the number of sides in the large shape in each of the first three pictures.

This is a difficult question because of the number of **distractors** (that is, changing features that are not part of the answer). These are the number of shapes, their shading and which way up they are (their orientation).

The only way to solve these tricky problems is to look at all the options one at a time without getting put off. There will be an answer – it might just need a bit of detective work to find it.

2 Numbers, shapes and relationships

28

Train

1 What is the number connection between these pairs of pictures?
(a)
(b)

Try

Look at the first two pictures and decide what they have in common. Then select the option from the five on the right that belongs to the same set. Circle the letter beneath the correct answer.

2 a b c d e
3 a b c d e

Test Test time: 2:00

4 a b c d e
5 a b c d e
6 a b c d e

Working with numbers

29

Test

Three timed questions to see what real test questions are like. *The last one will be difficult*, so if you can do this one, the skill is nearly mastered. Always complete the questions even if you don't manage them in the time. The practice is important.

- **End-of-chapter tests** give you a chance to try out the skills you have learned in a short test. These tests may include some different formats to see how you get on when tackling questions presented in slightly different ways. The test time given is for a *slow* test in Part 1 and an *average test* in Parts 2 and 3, although some of the most challenging tests you may encounter in your 11+ give you less time. Always time yourself to build up your speed.
  - Complete your first attempt in pencil or on a separate sheet of paper and aim for the test time given.
  - Complete the questions you don't finish in the time (but mark down which ones they are).
  - Go through the test again with a friend or parent and talk about the questions you found tricky.
  - Later on, have a second attempt and aim to work through all the questions within the time to practise working at speed.
- **11+ sample test** The test at the end of the book is based on a *difficult test* to be completed in a *challenging time*. A test like this could be set by an independent school. You may find some

of the questions hard and may also find it difficult to complete it within the time, but don't worry. This is a training test to build up your skills. Agree with your parents on a good time to take the test and set a timer going. Prepare for the test as if you are actually going to sit your 11+ (see 'Test day tips' below).

- Complete the test with a timer, in a quiet room. Note down how long it takes you, writing your answers in pencil.
- Mark the test using the answers at the back of the book.
- Go through the test again with a friend or parent and talk about the difficult questions.
- Have another go at the questions you found difficult and read the answers carefully to find out what to look for next time.
- If you didn't finish the test in the given time, have another attempt before moving on to other practice tests in the *Practice Papers* (see 'Continue your learning journey' on the next page).

- **Answers** to all the tests in this book can be found in the cut-out section beginning on page 111. Try not to look at the answers until you have attempted the questions yourself. Each answer has a full explanation so you can understand why you might have answered incorrectly.

Test day tips

Take time to prepare yourself the day before you go for the test: remember to take sharpened pencils, an eraser and a watch to time yourself (if you are allowed – there is usually a clock present in the exam room in most schools). Take a bottle of water in with you, if this is allowed, as this will help to keep you hydrated and improves your concentration levels. ... and don't forget to have breakfast before you go!

For parents

This book has been written to help both you and your child prepare for both Pre-Test and 11+ entrance exams.

The book doesn't assume that you will have any prior knowledge of Non-Verbal Reasoning tests and is designed to help you support your child with clear explanations for parents at the beginning of each chapter, which include:

- links to familiar areas of Maths using simple diagrams
- advice on how additional work in Maths can have a real impact on success in Non-Verbal Reasoning
- activities and games to enjoy with your child to support them in fun ways, involving other members of the family.

The teaching content is designed so that it can be tackled in simple steps. Setting aside time when your child can concentrate fully on one or two types of question when you are there to support them can help to make the experience manageable and enjoyable.

All the answer explanations tie into the teaching given. So if, for example, the question involves scaling and angles, the answer explanation will make this clear. This approach has been followed in order to support you and your child in reviewing questions they may have found challenging and so provide ideas for patterns to look out for when practising further questions.

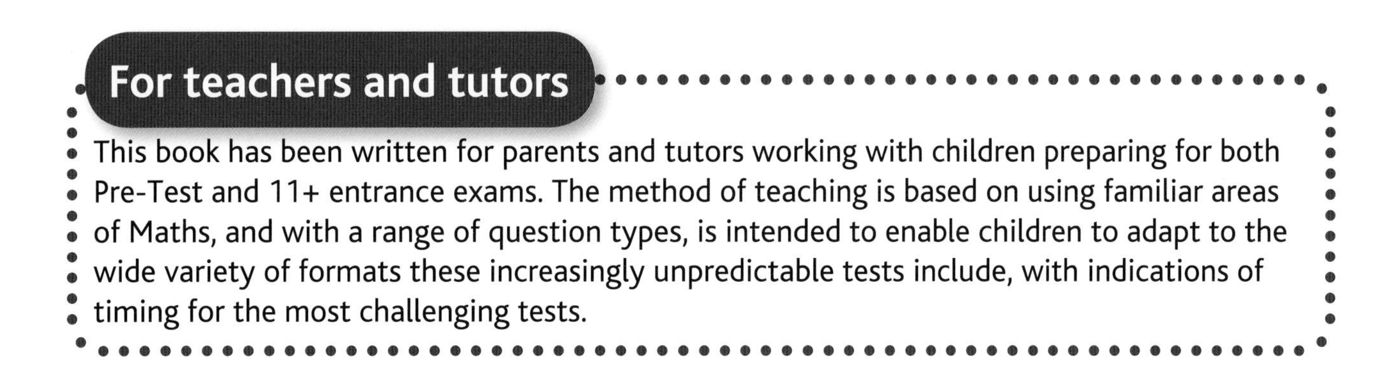

For teachers and tutors

This book has been written for parents and tutors working with children preparing for both Pre-Test and 11+ entrance exams. The method of teaching is based on using familiar areas of Maths, and with a range of question types, is intended to enable children to adapt to the wide variety of formats these increasingly unpredictable tests include, with indications of timing for the most challenging tests.

Continue your learning journey

When you have completed this *Study and Revision Guide*, you can carry on your learning right up until exam day with the following resources.

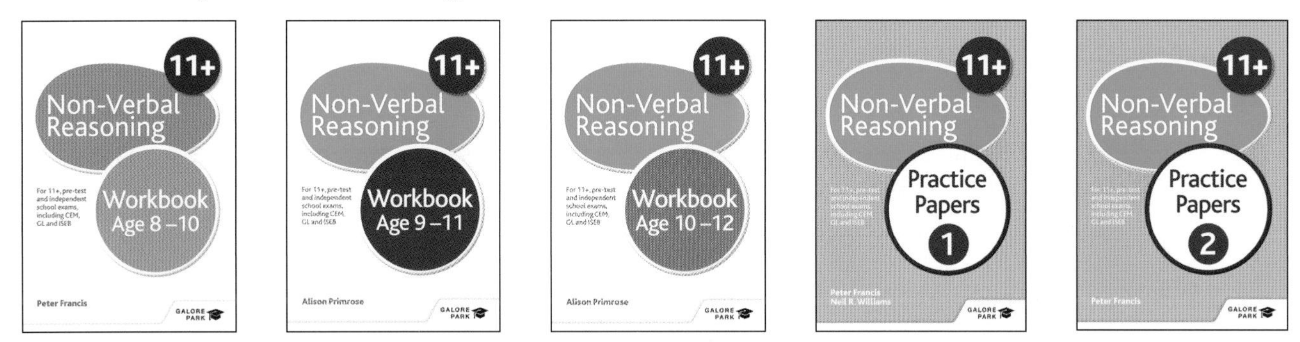

The *Workbooks* will develop your skills with over 160 questions to practise in each book. To prepare you for the exam, these books include even more question variations that you may encounter – the more you do, the better equipped for the tests you'll be.

- Age 8–10: Increase your familiarity with variations in the question types.
- Age 9–11: Experiment with further techniques to improve your accuracy.
- Age 10–12: Develop fast response times through consistent practice.

The *Practice Papers* (books 1 and 2) contain four training tests and nine model exam papers, replicating the different Pre-Test and 11+ exams you may encounter. They also include realistic test timings and fully explained answers for final test preparation. These papers are designed to improve your accuracy, speed and ability to deal with variations in question format under pressure.

Part 1: What is Non-Verbal Reasoning?

1 General themes and question types

Introduction

When you first look at a set of Non-Verbal Reasoning questions, you may find it difficult to know where to start. They may not seem to relate to anything you have encountered at school, so where do you begin?

You might be surprised to find out that Non-Verbal Reasoning questions are closely linked to areas of Maths. The learning ladders below show you how these unfamiliar questions are actually testing skills you already have, just in a slightly different way. Simpler Non-Verbal Reasoning skills appear at the bottom of the ladders, working towards the more challenging skills at the top.

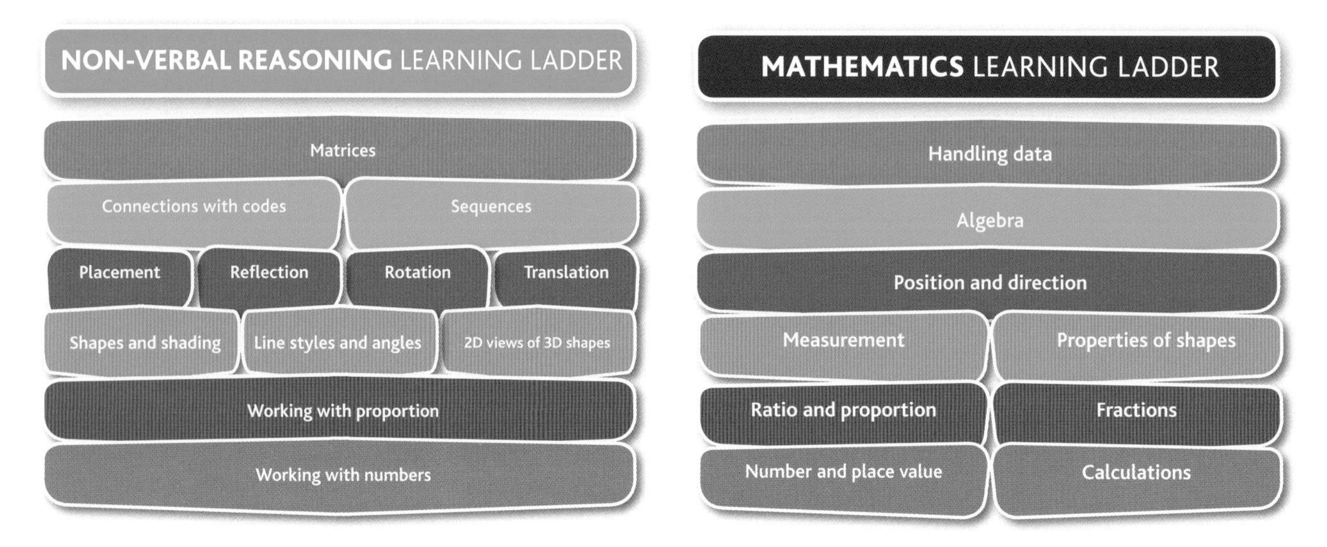

This section, Part 1, explains which areas of Maths can help you break down the questions into steps that are easy to follow. It then looks at a few of the different types of questions you might find them in.

In Parts 2 and 3, you will work from the simpler skills at the bottom of the learning ladder to the more challenging skills further up. You will reach the top of the ladder in Part 3!

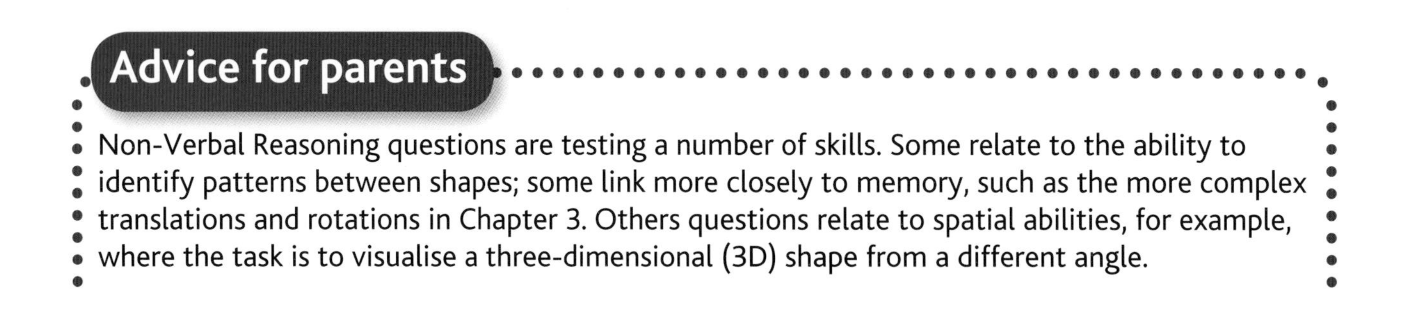

Advice for parents

Non-Verbal Reasoning questions are testing a number of skills. Some relate to the ability to identify patterns between shapes; some link more closely to memory, such as the more complex translations and rotations in Chapter 3. Others questions relate to spatial abilities, for example, where the task is to visualise a three-dimensional (3D) shape from a different angle.

Adults as well as children find some of these skills easier to master than others. Therefore, many strategies are provided in Parts 2 and 3 of this book to help your child improve on those areas that they find most challenging.

The test at the end of this chapter gives your child a feel for what the questions entail before they get involved in the detail in Chapter 2. Looking through the answers together afterwards will give you an idea about which particular skills your child found challenging and may need to spend a little more time on as they work through the book. This test is of a lower level of difficulty and slower timing than those in the following chapters to enable your child to answer some questions successfully and build confidence.

Thinking skills and games

The following games can help to develop skills in different aspects of Non-Verbal Reasoning. You will find further activities related to these skills at the beginning of each chapter.

Breakfast brain-boxes

Choose a large cereal packet to use for this game. Everybody at the table has one minute to look at a chosen side of the packet and memorise everything on it.

Take it in turns to stick opaque tape or sticky notes over an image or word. The next person must try to remember the missing image or word. Score a point for a correctly remembered feature. The person with the most points is the winner.

You can do the same activity with the headlines on a newspaper page.

Emoji flips

This is a game for two or more players.

Choose an emoji from a mobile device or computer. The emoji should be symmetrical, such as a Smiling Face or Heart Eyes Face. Each player should first draw the image before beginning the game. The challenge is then to draw a reflection across a horizontal line (so that the face appears to be standing on its head).

Draw two more symmetrical images in the same way and then move on to three asymmetrical images (such as a Wink or a Smug Face) and draw vertical reflections.

The person drawing the correct image first gains two points. Other players who draw a correct image get one point. Players gain no points if the image is incorrect in any way, such as the eyes being the wrong way up. Use a mirror if there is any doubt about whether the image is correct. The game is won by the player with the most points after six images are drawn.

Setting a maximum time per round makes the game more difficult.

Shape spotting

Before you go for a walk or drive choose a 2D or 3D shape to look out for, such as a hexagon. The first person to spot the chosen shape wins a point (this could be a hexagon on a flower, on a sign, on a leaded window or on a football) and then chooses the next shape. The person with the most points at the end of the trip wins the game.

What you can expect to see

Most Non-Verbal Reasoning questions ask you to find features that are either the same or different in a set of pictures. Although you will come across many different types of questions with the pictures presented in a variety of ways, the purpose of each type of question is the same – to find out whether you can spot these similarities and differences.

Understanding the elements

Non-Verbal Reasoning questions are made up of several elements. You will see a variety of:

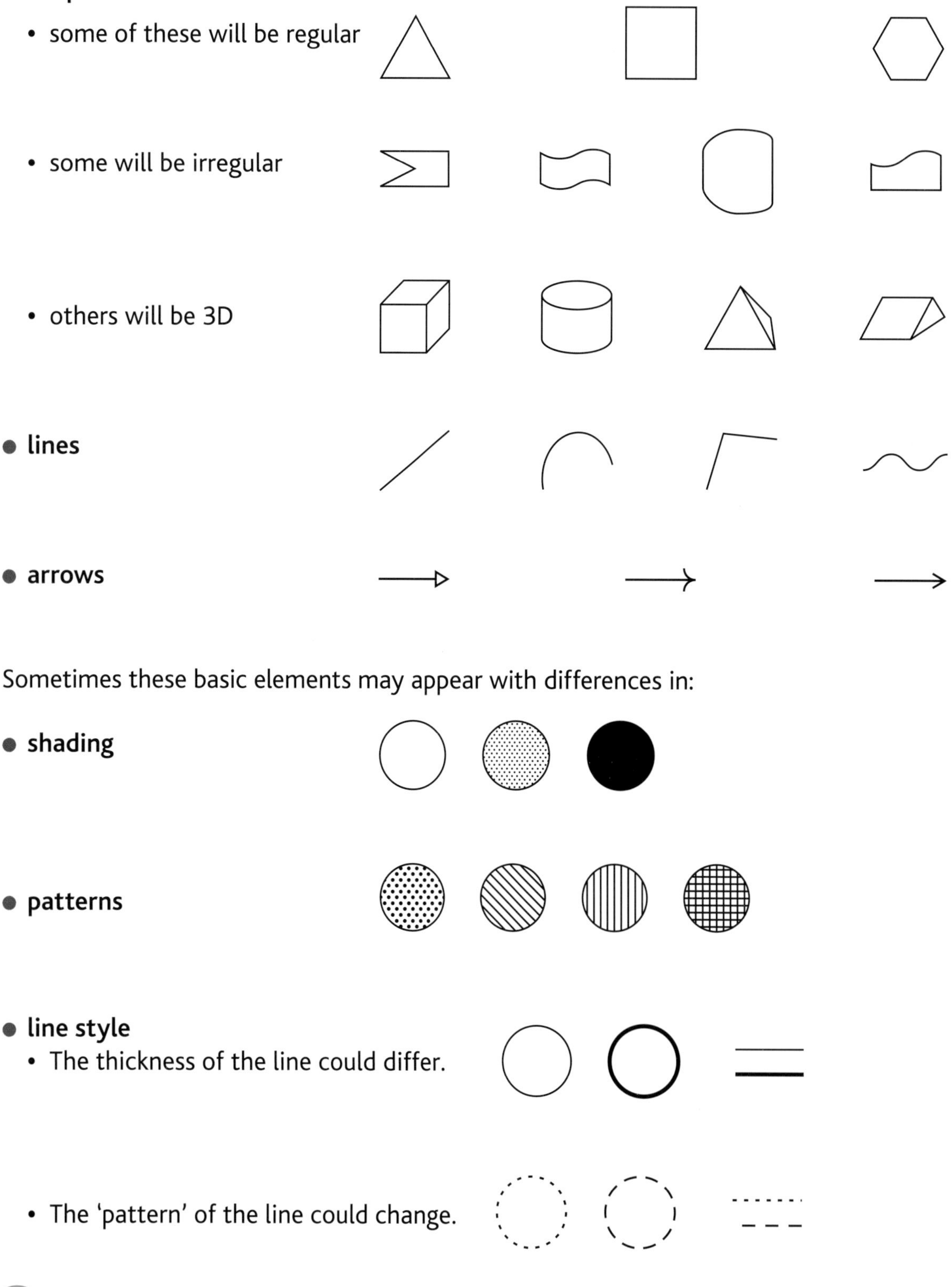

- **shapes**
  - some of these will be regular
  - some will be irregular
  - others will be 3D
- **lines**
- **arrows**

Sometimes these basic elements may appear with differences in:

- **shading**
- **patterns**
- **line style**
  - The thickness of the line could differ.
  - The 'pattern' of the line could change.

Recognising the patterns

Every test will include a different combination of elements and styles. Part 2 of this book looks at how these patterns change.

In Chapter 2 you will explore:

- **number**
- **scale and proportion:** shapes can change in size or a fraction of a shape may be shaded.
- **angle**

In Chapter 3 you will explore:

- **position and direction:** the shapes appear to point in a certain direction or to overlap each other, as if you are looking at a 3D image.
- **rotation:** the triangle is moving clockwise within the square and also rotating clockwise by 90° each time.
- **reflection:** the shape in the first box is reflected horizontally while the shape in the second box is reflected vertically.
- **translation:** the two triangles in the first box move together in the second box, but otherwise do not change position.

Train

1 What has changed between the first and second images in each question?

(a) 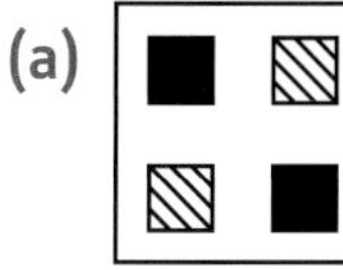 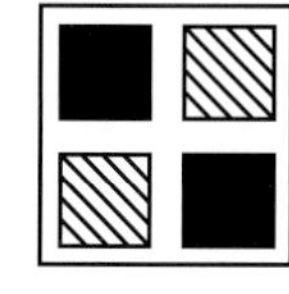

(b) 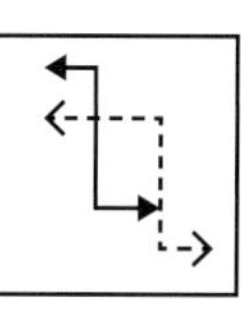

______________________ ______________________

2 Compare these pictures.

(a) Write one thing they have in common. ______________________

(b) Write two things that are different. (i) ______________________

(ii) ______________________

Working out what is important

When faced with a question containing many different elements, you may have to work through more than one step find the right answer. Luckily, there are a number of techniques you can follow to help you.

Looking at the patterns

The simplest Non-Verbal Reasoning questions ask you to look at one feature to find a link. This may be spotting something that is the same or something that is different as you did in the *Train* activity on page 13.

More complicated questions generally include more elements so you have to look at more than one feature.

- Look at the example on the right. Think about the different elements that might change. See if you can find *three* differences between the two pictures before reading the following explanation.

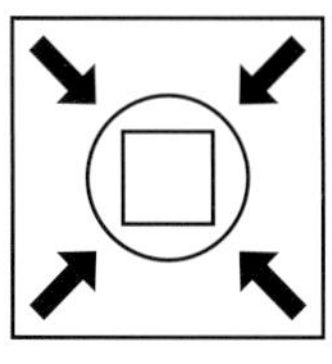

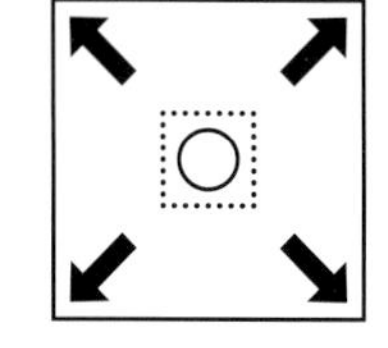

  - The arrows in the corners are pointing inwards in the first picture but have **rotated 180°** to point outwards in the second picture.
  - The circle is **scaled down** to fit inside the square.
  - The square stays the same size, but its **line style changes** from a solid line to a dotted line.
- Every element in the picture has changed in this example.

Identifying elements that do not matter

Many questions also involve elements that are not important. These random elements are called 'distractors' and are put in to make it more difficult to spot the patterns.

- Look at the example on the right. Two elements change from the first picture to the second picture. Try to work out what they are before reading the explanation.

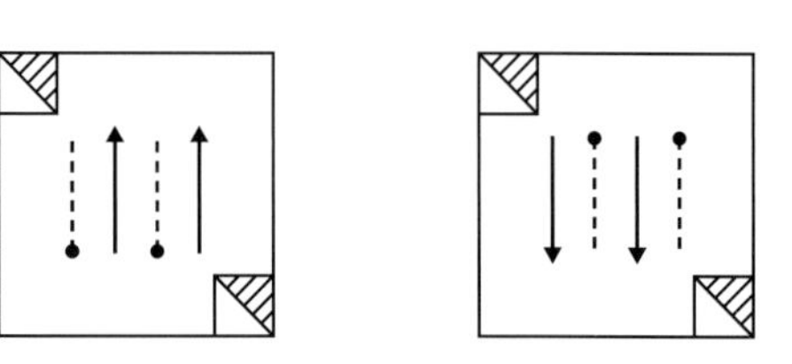

  - The two different line styles swap **position**.
  - The arrows also change **direction**.
- You will probably have also have looked at the squares in the corners and noticed that **they do not change**. These are *distractors*. They do not help you get the answer, but you look at them simply because they are there. It is very easy to get tripped up in a Non-Verbal Reasoning question by distractors as they create false trails. Expect to see them and be ready to ignore them!

Spotting connected elements

When you have looked at all the clues in a question and cannot seem to see a pattern, it is possible that two of the elements have a connection.

- Look at the example on the right. Try to work out what the connection might be before reading the explanation. It is quite hard to spot since there are three distractors to lead you in the wrong direction.

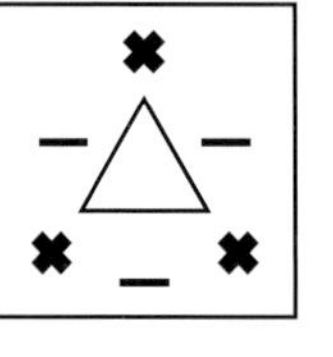

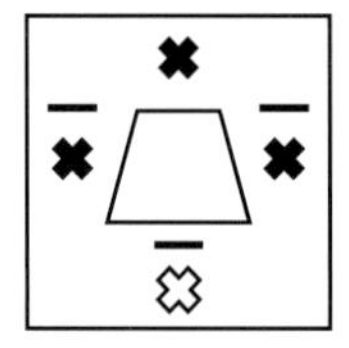

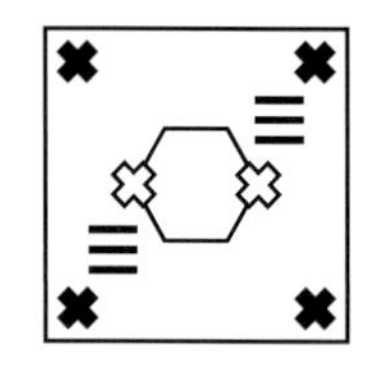

- At first the **number of short lines** looks like the common feature. The first two pictures both include three but the third picture includes six. This does not make sense so a random pattern like this is usually a distractor.
- The **number of black crosses** and the number of **white crosses** also follow a random number pattern and so must be distractors too.
- The **position of the crosses and lines** look as if they are in a pattern, but some are near vertices (corners), some near the sides and others move about. This is another distractor.

- The one feature we have not looked at yet is the **total number of crosses**: there are three in the first picture, four in the second and six in the third. Now look at the large shapes in each picture. The first has three sides (and vertices), the second four and the third six. So the link between the three pictures is that the **total number of crosses equals the number of sides of the shape**.

Remember: Random elements are almost always distractors!

Train

1 List four things that have changed between the first and second pictures.

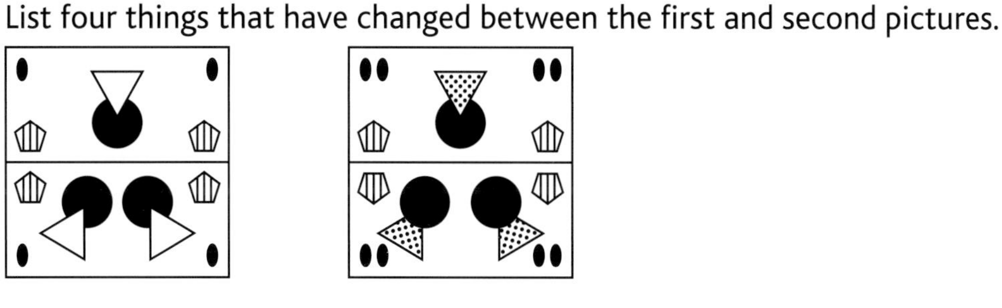

(i) ____________________

(ii) ____________________

(iii) ____________________

(iv) ____________________

2 Look closely at the four pictures below. Two of the elements are connected in some way.

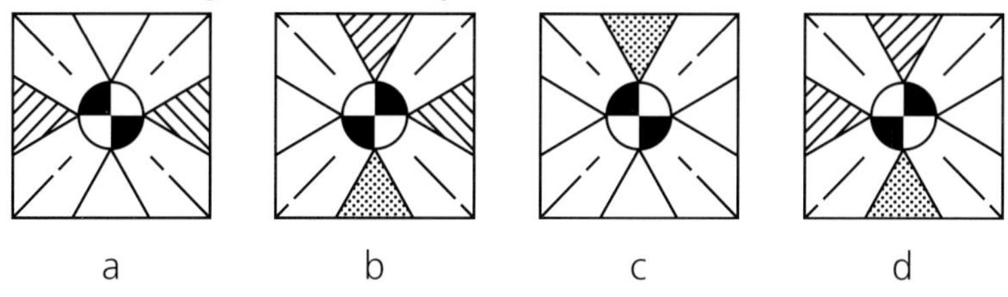

(a) Describe one random distractor in these pictures. ____________________

____

(b) What is the connection between two of the elements in the pictures?

____

____

Questions where the picture changes

These types of questions ask you to look at a number of different pictures to spot **similarities** and **differences**. We have also included some questions that ask you to imagine a **2D shape in 3D** (and the other way around) – the picture still changes, just in a different way. We look more closely at these question types in Chapters 2 and 3.

Most unlike

Look at these pictures. Identify the one that is most unlike the others. Circle the letter beneath the correct answer.

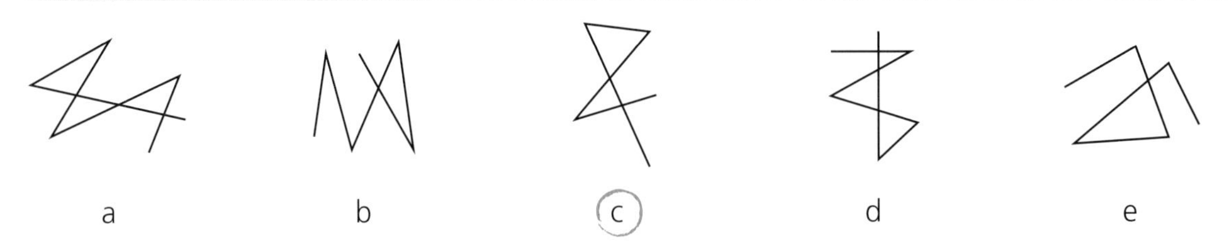

- These questions are often the first type you will come across in Non-Verbal Reasoning tests as they are quite simple to understand. You will usually see five options to choose from. All but one of the options will have features in common. The answer option will have **one feature** that is **different**.

Matching features

Look at the first two pictures and decide what they have in common. Then select the option from the five on the right that belongs in the same set. Circle the letter beneath the correct answer.

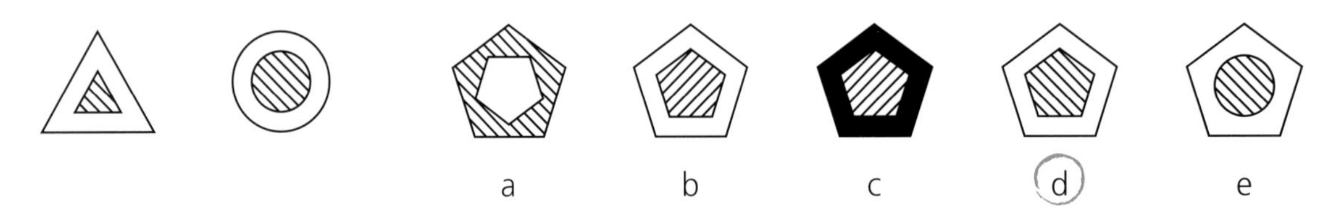

- Before looking at the answer options in these questions, you first need to work out what links the two pictures given on the left. They may look quite different, but they will always have **something in common**. Only one of the answer options will have this common feature.

Applying changes

Look at the two pictures on the left connected by an arrow. Decide how the first picture has been changed to create the second. Now apply the same rule to the third picture and circle the letter beneath the correct answer.

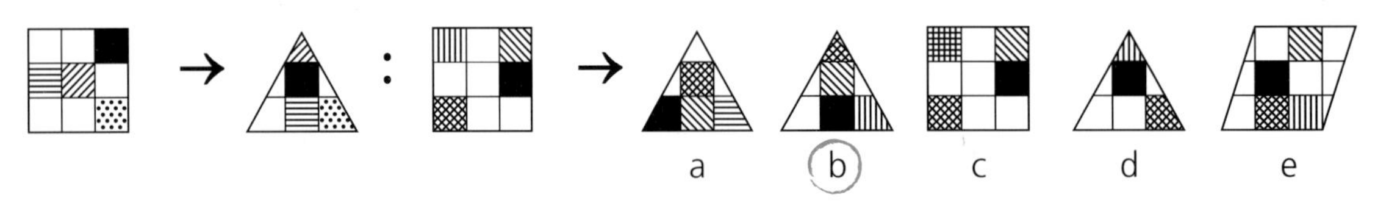

- Rather than looking for something in common in these questions, you need to spot the **feature or features that have changed**.

- Here the most obvious difference between the first and the second picture is the change of **shape** from a square to a triangle. But there are three triangles among the answer options, so there must be at least one more feature.
- Look at the **shading** in the first picture. There are four different patterns: vertical lines, solid black, diagonal lines and dots. Although the position of these patterns is different in the second picture, the **combination of patterns** remains unchanged.
- Work through the triangles in the answer options to see which one matches the shading pattern of the third picture. The only triangle with the same shading pattern is **b**.

Matching a net to a cube

Find the cube that can be made from the net shown on the left. Circle the letter beneath the correct answer.

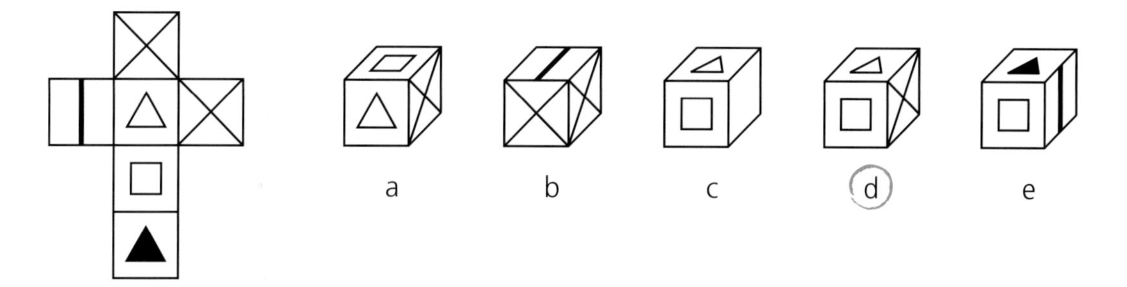

- These 3D questions also involve looking for patterns, but this time you have to imagine how the pattern matches up if the net is folded into a 3D shape. Variations on these questions may use different 3D shapes or begin with a 3D shape and ask you to match it up to a net. You can work out all these questions using the techniques explained on page 36.

Matching a 3D picture to a plan view

Other questions ask you to look at a 3D picture and imagine what it would look like when viewed from above. You are then given a number of possible answers to choose from.

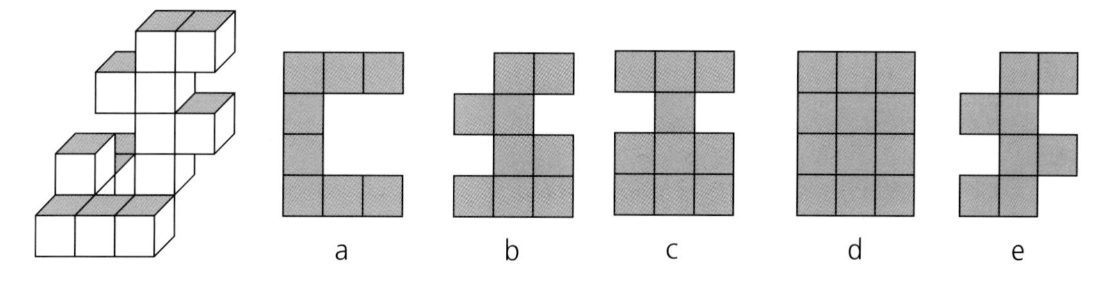

- Techniques to help you work out the correct answer to these questions are explained on page 40.

Train

1 Look at the *Most unlike* question on page 16. Why is option **c** the correct answer?

____

2 Look at the *Matching features* question on page 16. Why is option **d** the correct answer?

____

Questions where the picture moves

Most of the questions earlier in this chapter have looked at pictures that change between the question and answer options. For all the question types in this section, you need to spot the **same** picture from a **different viewpoint**. This tests your Maths skills of position and direction, rotation, reflection and translation. We look more closely at these question types in Chapter 3.

Following the folds

The square given at the beginning is folded in the way indicated by the arrows, and then holes are punched where shown on the final diagram. Identify the answer option that shows what the square would look like when it is unfolded. Circle the letter beneath the correct answer.

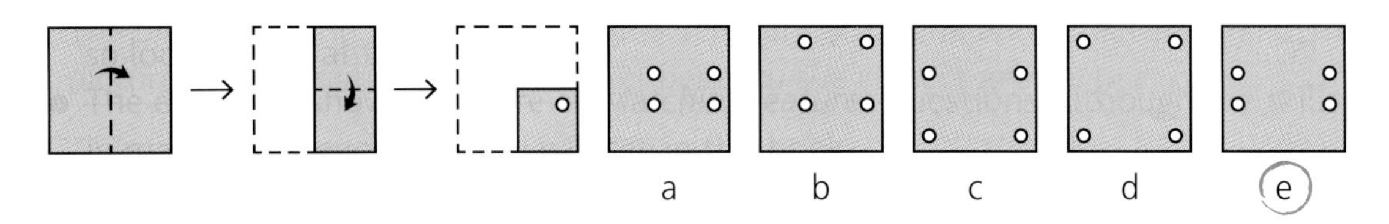

- Questions involving folds are really looking at reflections and symmetry. Each fold is actually a line of symmetry and it is important to **work backwards** from the last fold to the first fold to find out where the holes will appear.
- A variation on this question type shows an unfolded shape and asks you to imagine it folded up. This variation is unusual to find but no more difficult to answer.

Matching a reflected or rotated 2D picture

The picture on the left is reflected in a vertical mirror line and is represented by one of the pictures on the right. Circle the letter beneath the correct answer.

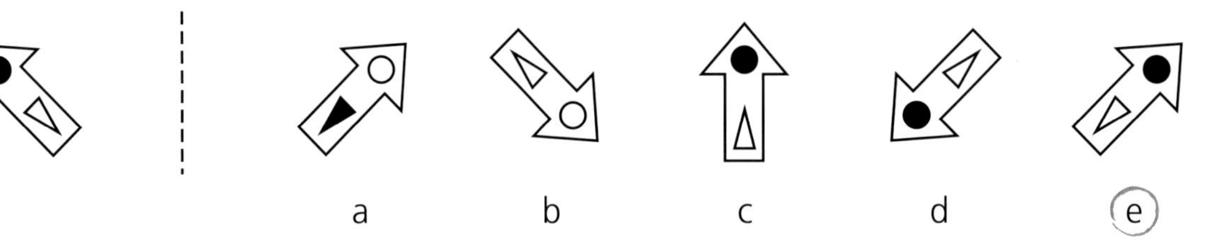

- There are many variations of this question type: some ask you to *reflect* a picture (as shown in this example) while others ask you to *rotate* it. You may even be told how many degrees the picture has been rotated.
- Working through the elements in the source picture one at a time can help you to solve all of these questions quite simply. In this example, only options **c**, **d** and **e** have a black circle like the one in the arrow on the left. This means that you can ignore options **a** and **b**. Now, if you imagine holding a mirror up to the right of the source picture, the only possible answer is **e**.
- You will find out more about *Rotation* and *Reflection* questions on pages 50–55.

Choosing a rotated 3D picture

One of the two pictures on the left is a rotated version of the 3D diagram on the right. Circle the letter beneath the correct answer.

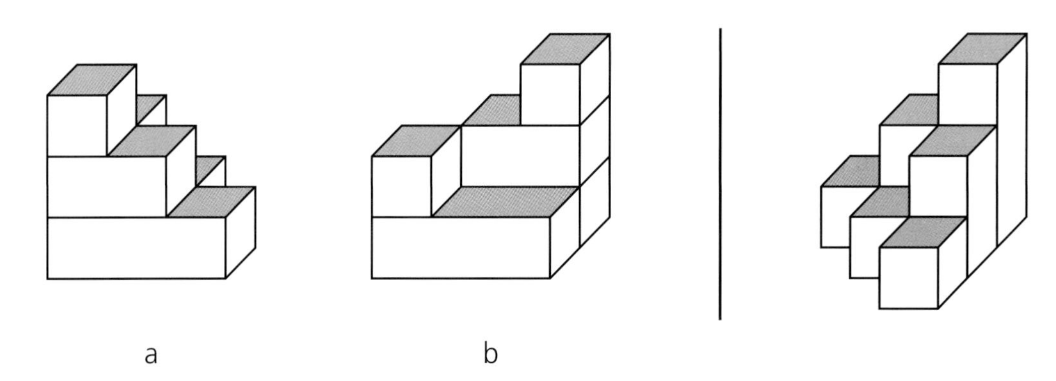

- These questions are presented in a variety of ways – sometimes with a group of five pictures to be matched up with a further five rotated pictures. You will find out more about how to tackle these questions on pages 56–57.

Translating and combining shapes

The small shape on the left can be found in one of the pictures on the right. It might be made up of one or more pieces. Circle the letter beneath the correct answer.

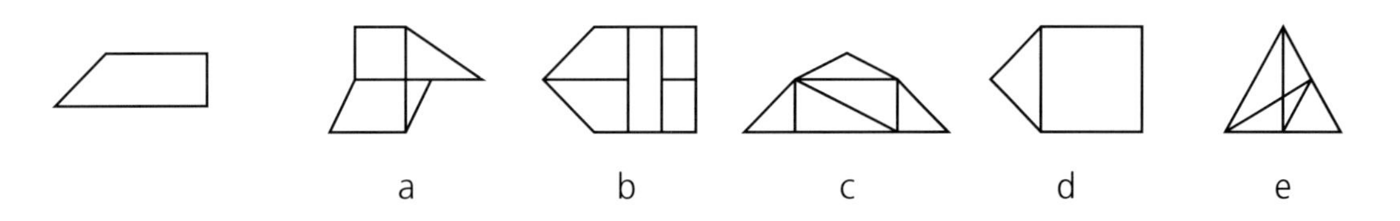

- Several types of questions involve translations and may involve either 2D or 3D shapes.
- These questions often just ask you to find exactly the same outline in the options you are given, although sometimes you may have to reflect or rotate an image as well. You will find out more about translating shapes on pages 58–62.

Train

1 Look at the *Choosing a rotated 3D picture* question above. Decide which is the correct rotation of the picture on the right and write the answer option on the line provided here.

______

2 Look at the *Translating and combining shapes* question above. Trace around the shape on the left and overlay it on the answer options until you find the correct answer. Write your answer on the line provided here.

______

Questions involving algebra

You may have been asked to work out problems in Maths where a letter stands for a number in a calculation. Question types involving codes use a similar idea, with each code **letter** linking to a **feature** in the pictures you have been given. Other question types in this group are based on sequences, where you are asked to follow a changing pattern. You will look more closely at these question types in Chapters 4 and 5.

Connections with codes

Each letter represents an individual feature in the picture next to it. Work out which feature is represented by each letter. Apply the code to the picture in the box and circle the letter beneath the correct answer code.

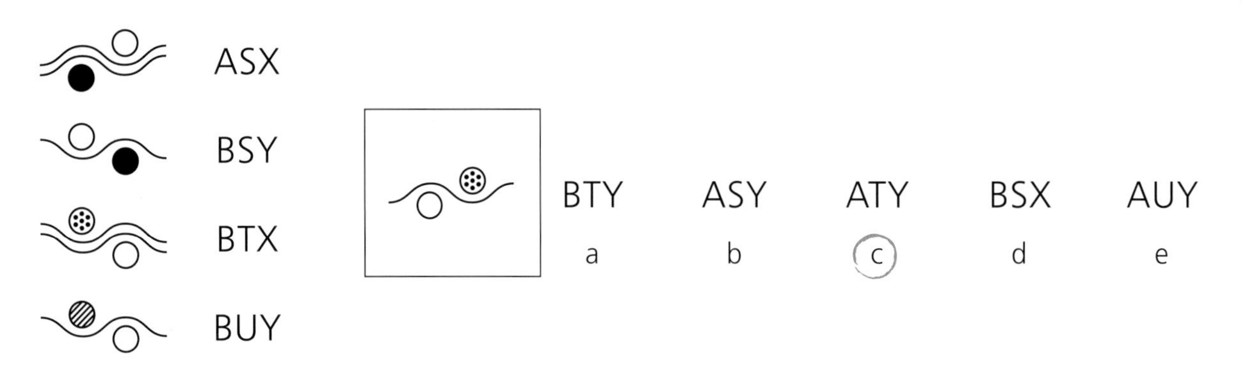

- You may see *Connections with codes* questions that give you one, two or three code letters to work out. These questions are solved in two steps: you first need to work out what each letter stands for and, secondly, apply these codes to a new shape, before choosing the matching answer option.
- Take each letter in turn. Look at the first letter in the example above.
  - You will see one picture with the code letter A and three with the code letter B. This means that the second, third and fourth pictures have something in common (represented by the letter B) and that the first picture is the odd one out.
  - The curved line goes down and then up in the second, third and fourth pictures but up and then down in the first picture. This means that the first code stands for the **shape of the line**.
- Now move on to the second and third code letters and work out the features they represent in the same way.
  - The first and second pictures should have something in common (represented by the letter S), with the third and fourth pictures both being different (represented by the letters T and U). The first and second pictures both have a solid black circle; the third picture has a circle filled with a dotted pattern; the fourth picture has a circle with a diagonal pattern. Therefore the second letter must stand for the **shading**.
  - The only two other features of the pictures are the double curved line and the position of the shaded circles. Since the third letter of the code runs XYXY and the **number of curved lines** follow this alternate rule too, this must be the correct feature for the final code.
  - The position of the shaded circles is a distractor.

Sequences

The boxes on the left show a pattern that is arranged in a sequence. Choose the answer option that completes the sequence when inserted in the blank box. Circle the letter beneath the correct answer.

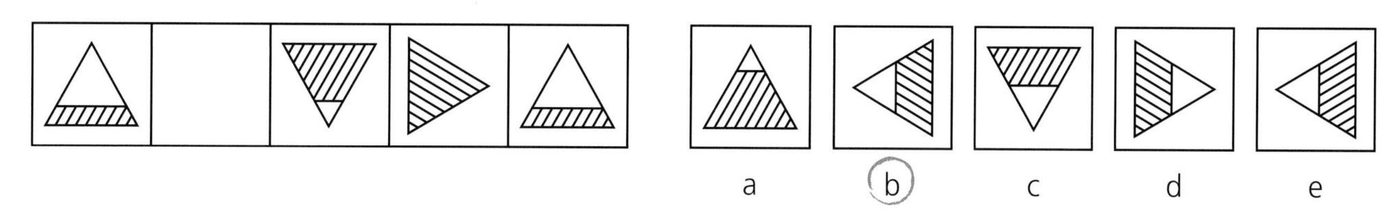

- *Sequences* questions ask you to **follow a pattern** or patterns. These sequences generally have four or five pictures in a row with one more to work out. Other styles of sequence have an extra row and are worked out in a similar way.
- In the example above the sequence involves two patterns that you need to follow to find the correct answer.
  - The triangle rotates 90° anticlockwise moving from left to right across the sequence. Following this rule, the triangle in the missing box will 'point' to the left, so options **b** and **e** are therefore possible answers (**d** is rotating clockwise).
  - The second pattern is the change of shading – this is a repeating pattern every four boxes. Moving left to right, the shading of the triangle increases gradually, before returning to the original shading in box five. So, in the second picture, the shading will come approximately halfway up the triangle. Options **b** and **e** are both still possible.
  - Options **b** and **e** look the same but, when you look closer, the direction of the shading is different. Imagine how the shading will work if you turn the triangle in the first box to the left. Turn the page if you find that easier. The correct answer is **b**.

Train

1 Look at the *Connections with codes* question on page 20. Draw a picture for the code BUX.

2 Imagine an extra box is added to the right-hand side of the *Sequences* question above. Draw the picture that would be in this box.

Questions involving matrices

Matrices look a bit like number grids, but do not contain any numbers! These questions are some of the most challenging you will come across because the patterns can run in different directions across and around the grids. You will find out more about these questions in Chapter 6.

Standard grids

- Most *Matrices* questions use a square grid although the number of boxes can change. The most common *Matrices* questions work in a 2×2 or 3×3 grid.
- In some questions the shapes will be enclosed in boxes but in others they are not in boxes. Both kinds of matrices work in exactly the same way so do not be put off by differences in the way they look.

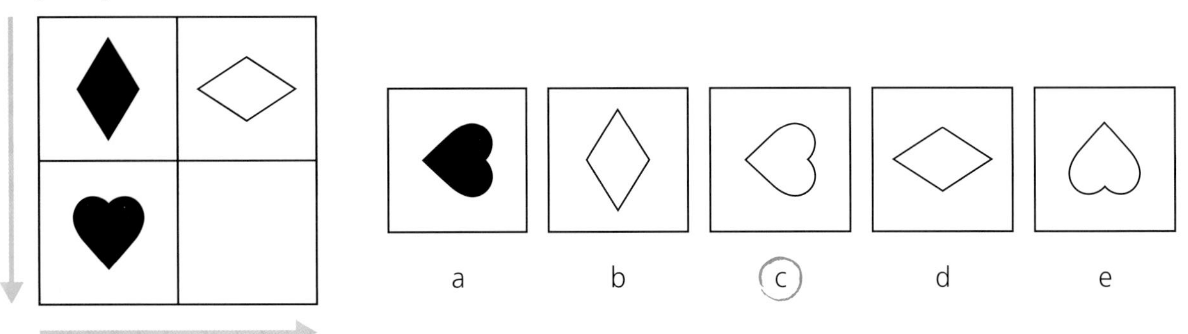

- 2×2 grids are the easiest to follow because there are only four boxes to look at and the patterns are nearly always **across and down**. Sometimes more than one change occurs between boxes.
  - First look at the columns: the shading is the same in each column (black in column 1, white in column 2); the shape rotates 90° between the columns.
  - Now look at the rows: the shape is the same within each row (diamond in row 1, heart in row 2), so the answer is **c**.

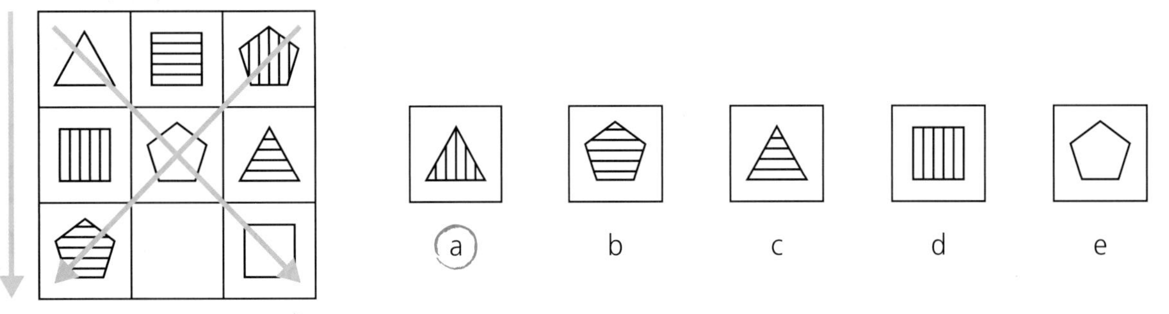

- 3×3 grids also work from top to bottom and left to right but can involve **diagonal patterns** as well.
  - Begin by looking down the columns: there is a different shape in each row, and each is shaded differently – either white, vertical or horizontal stripes.
  - Now look across the rows: the same rule applies, although, as with the columns, the pattern appears random. As there is no obvious rule, check the diagonals.
  - Look diagonally from top left to bottom right – the three shapes are all white. This is the most obvious diagonal, but there are two more to look at, working in this direction.
  - Now look diagonally from the middle box in the top row to the right-hand box on the middle row – this is the second diagonal. The square and the triangle have the same shading. The third diagonal is from the box on the left of the middle row to the empty box.

- However, the pattern of shapes still appears random so check the diagonal running in the other direction – top right to bottom left. All the central shapes are hexagons. Check the diagonal from the middle box in the top row to the left-hand box on the middle row. Both shapes are squares, so the empty box must be a triangle with vertical stripes to complete the grid.

Variations on standard grids

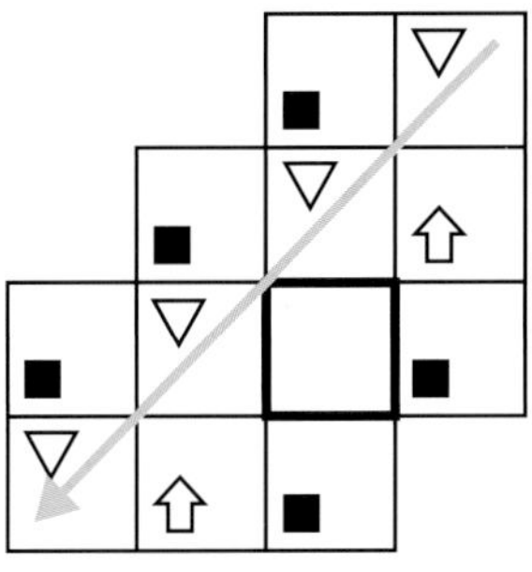

- The boxes in other *Matrices* questions may be organised in different ways – although you might be surprised to hear they are generally easier to solve than standard grids.
- The patterns work across, down and in diagonals in the same way. However, as there are often more boxes in a row, it can be easier to spot patterns. The example here has a simple diagonal pattern, so the missing shape is an upward-pointing white arrow.

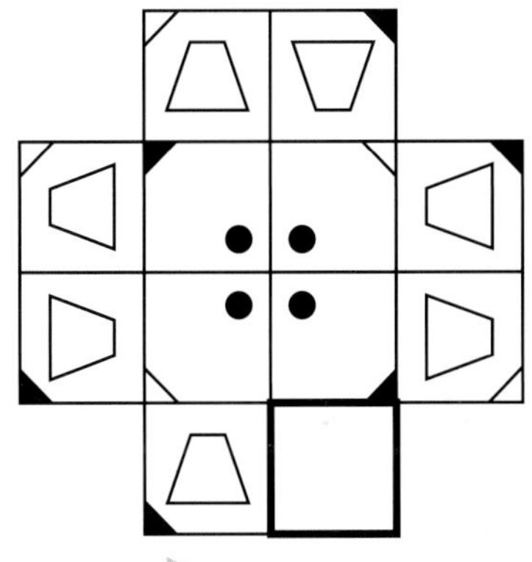

- Now look carefully at this cross-shaped grid. The pairs of boxes around the edge follow a straightforward pattern. You will see that this is consistent so you can actually ignore the central section of the grid completely when working out the answer. This kind of shortcut is not uncommon in these grids, so look carefully at the different sections to see if you can save yourself some work!

Isometric grids

- These grids use triangles or hexagons instead of square boxes. Owing to the shape of the grids, the patterns generally appear to work around a circle. This means that they create a continuous pattern that often works as a sequence. This grid is a good example containing a sequence of three elements: reflection, shading and number.

Train

1 Look at the cross-shaped grid in *Variations on standard grids* above. Draw the missing image in the box.

2 Look at the hexagonal grid in *Isometric grids* above. Draw the missing image in the triangle.

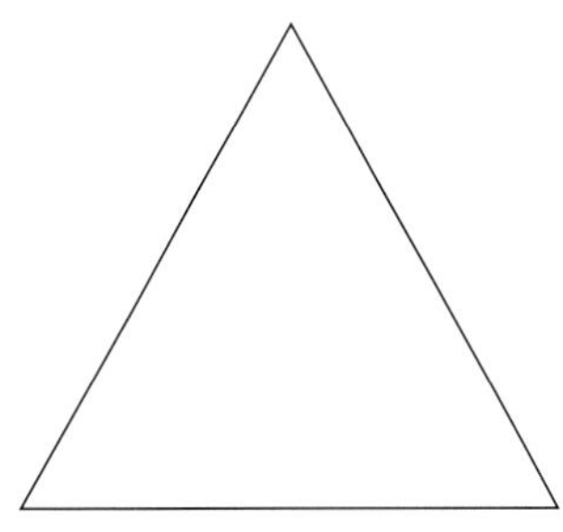

Now take the following short test to see how you get on with some real Non-Verbal Reasoning questions. Remember that these questions can take different forms, so watch out for variations in the questions you have already seen and read the questions carefully. Good luck!

Test 1

Test time: 06:00

Complete this test in the time given above. Each question is worth one mark.

Look at these pictures. Identify the one most unlike the others. Circle the letter beneath the correct answer. Example:

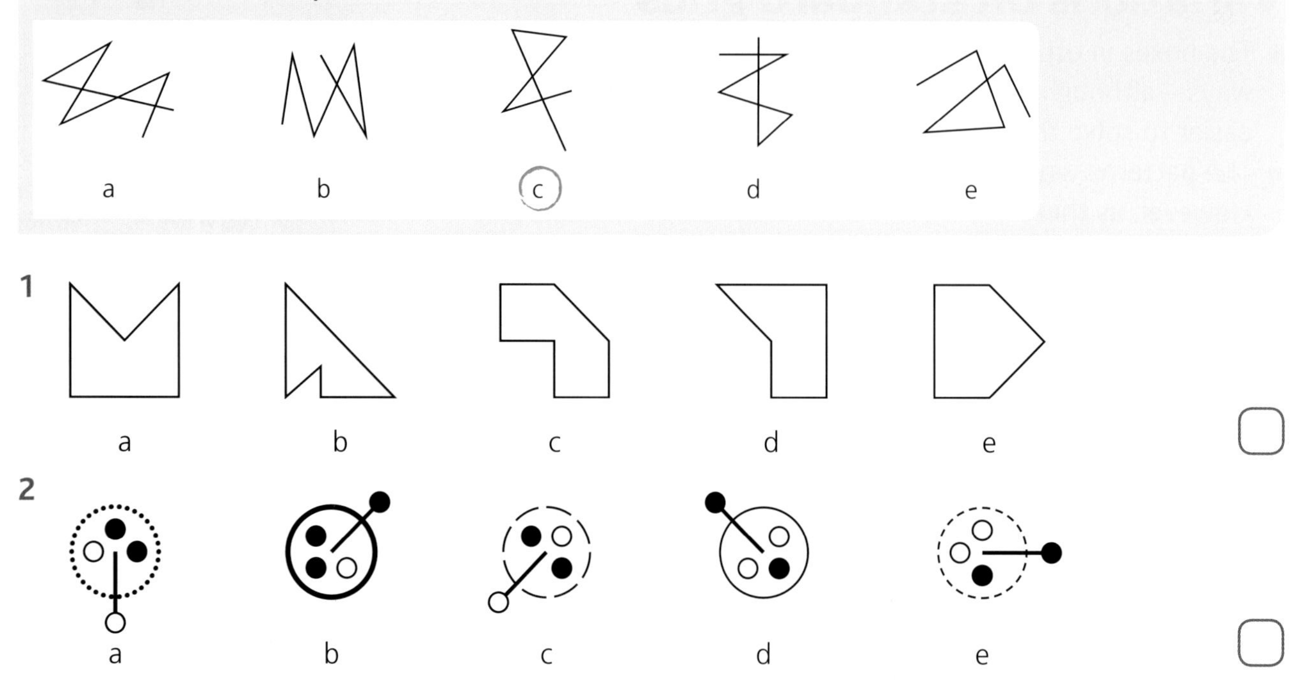

Look at the two pictures on the left connected by an arrow. Decide how the first picture has been changed to create the second. Now apply the same rule to the third picture and circle the letter beneath the correct answer. Example:

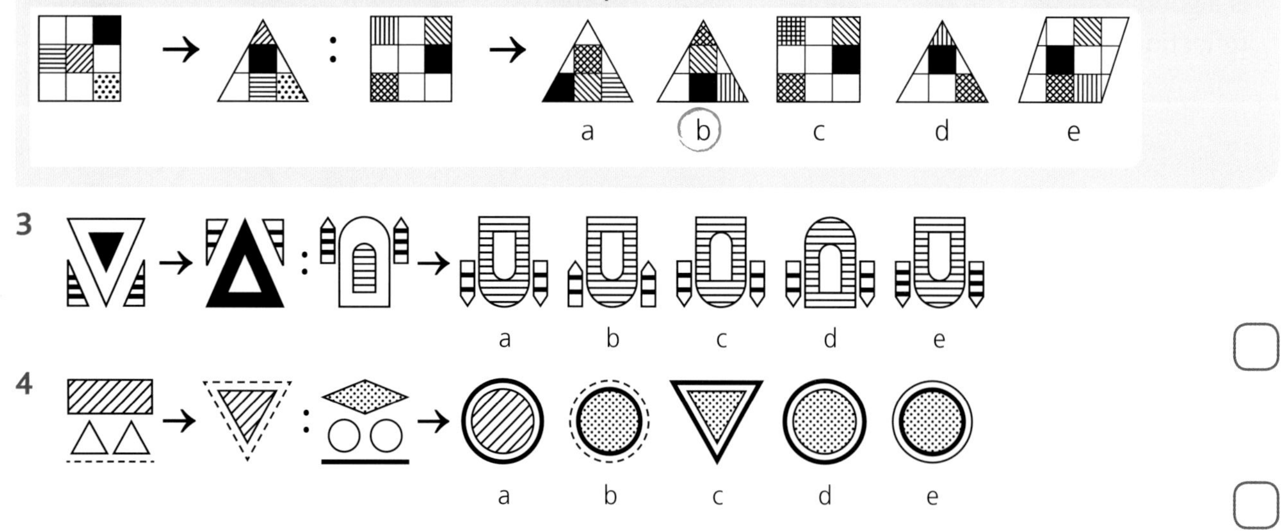

The picture on the left is reflected in a vertical mirror line and is represented by one of the pictures on the right. Circle the letter beneath the correct answer. Example:

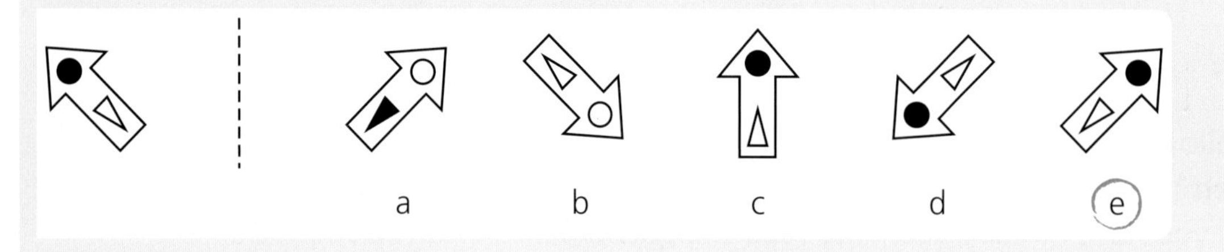

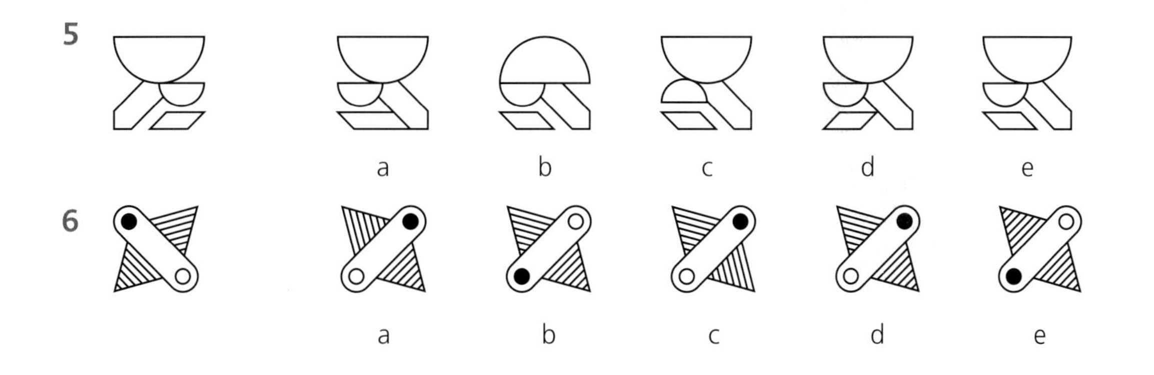

One group of separate blocks has been joined together to make the pattern of blocks shown on the left. Some of the blocks may have been rotated. Circle the letter beneath the blocks that make up the pattern. Example:

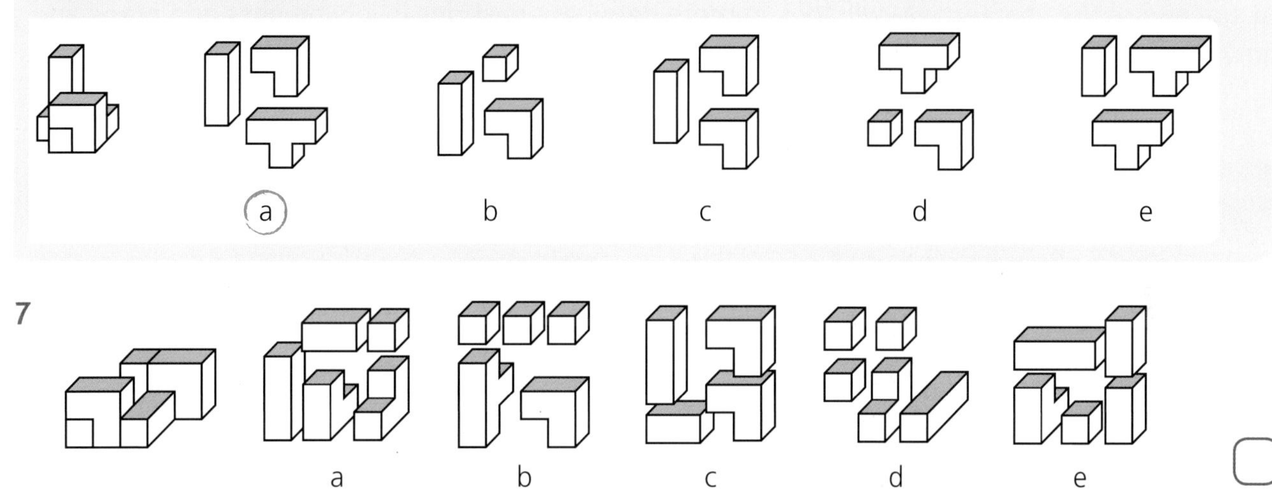

One of the options on the right completes the pattern in the grid on the left. Circle the letter beneath the correct answer. Example:

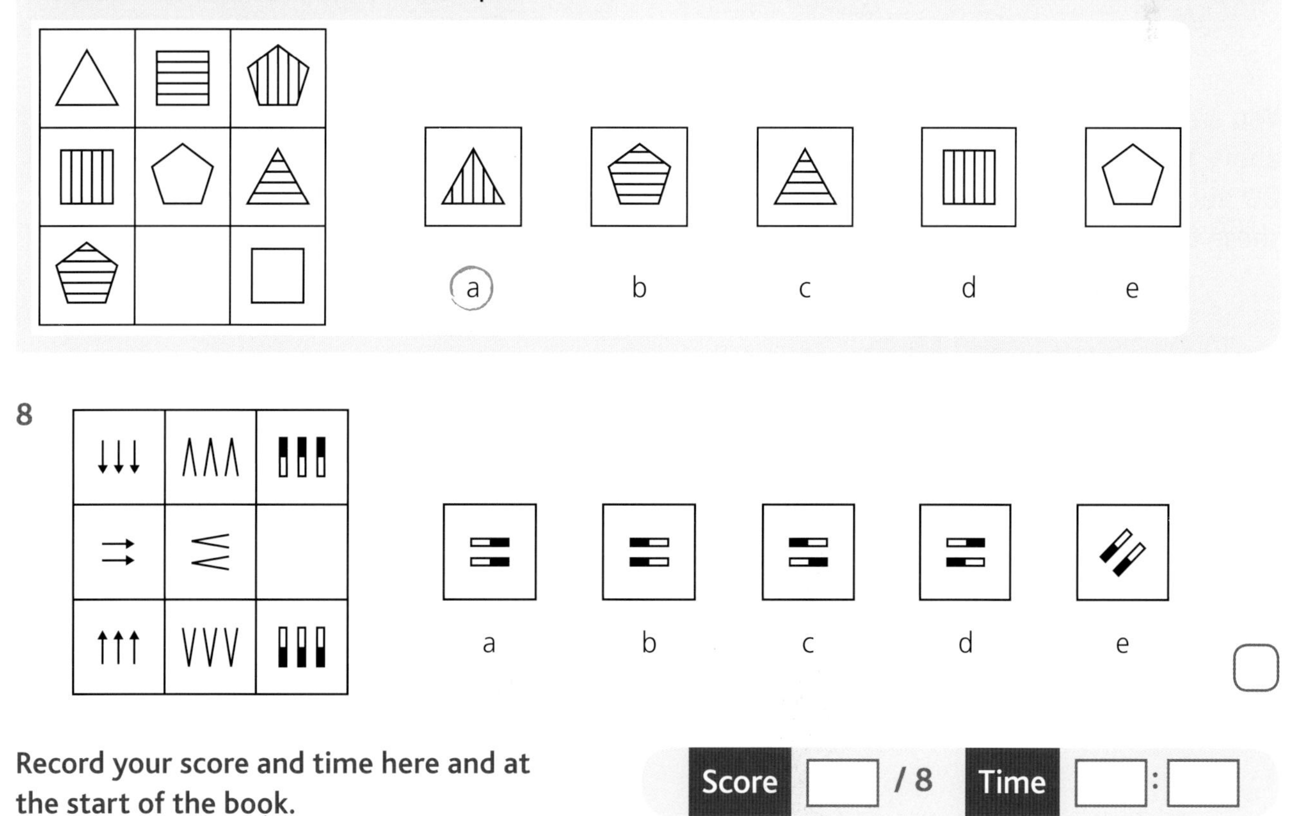

Record your score and time here and at the start of the book.

Score ☐ **/ 8** **Time** ☐ : ☐

Part 2: Skills for solving questions

2 Numbers, shapes and relationships

Introduction

In Part 1 you looked at some of the types of questions that appear in Non-Verbal Reasoning tests. In Part 2 you will look at the different features that can be found in these questions.

In this chapter, you will find out about the different kinds of **shapes**, **lines**, **shading** and **patterns** that are common to many questions.

At the end of the chapter, you will look in more detail at questions involving **nets** and **plans** and how to relate them to their **3D shapes**.

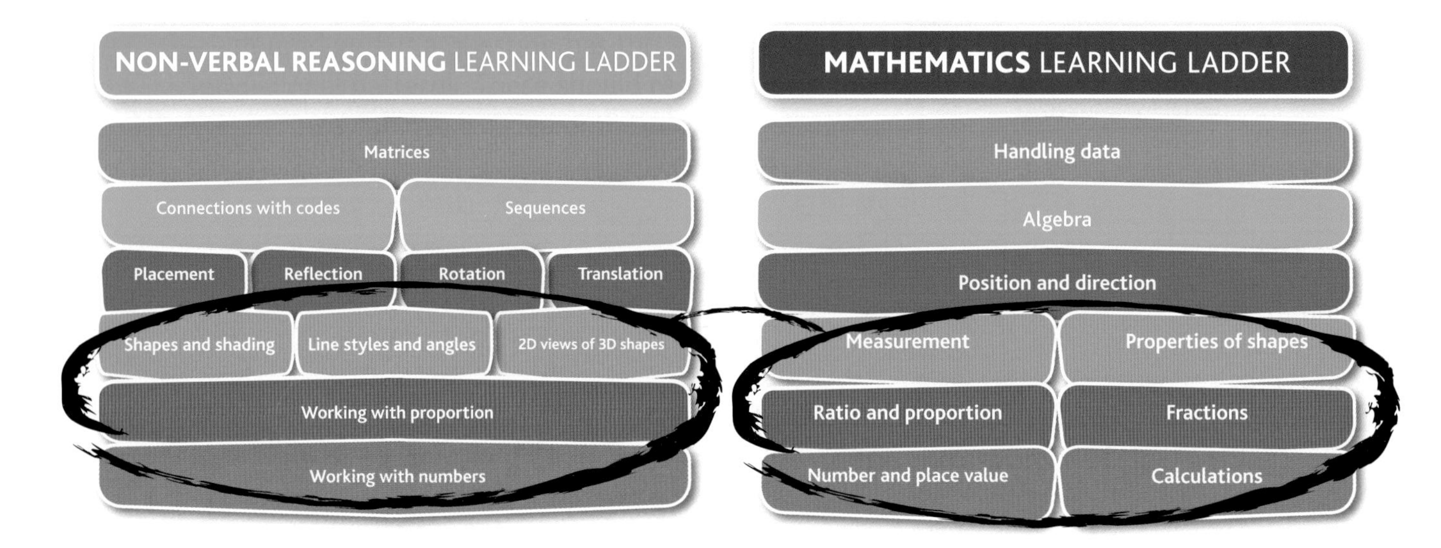

You are now going to climb the first three steps of the learning ladders. Compare the two ladders above to see how the Non-Verbal Reasoning topics of **numbers, proportion, shapes, angles** and **2D and 3D shapes** link to the different areas of Maths. All the questions in this chapter relate to these three steps, so let's get climbing!

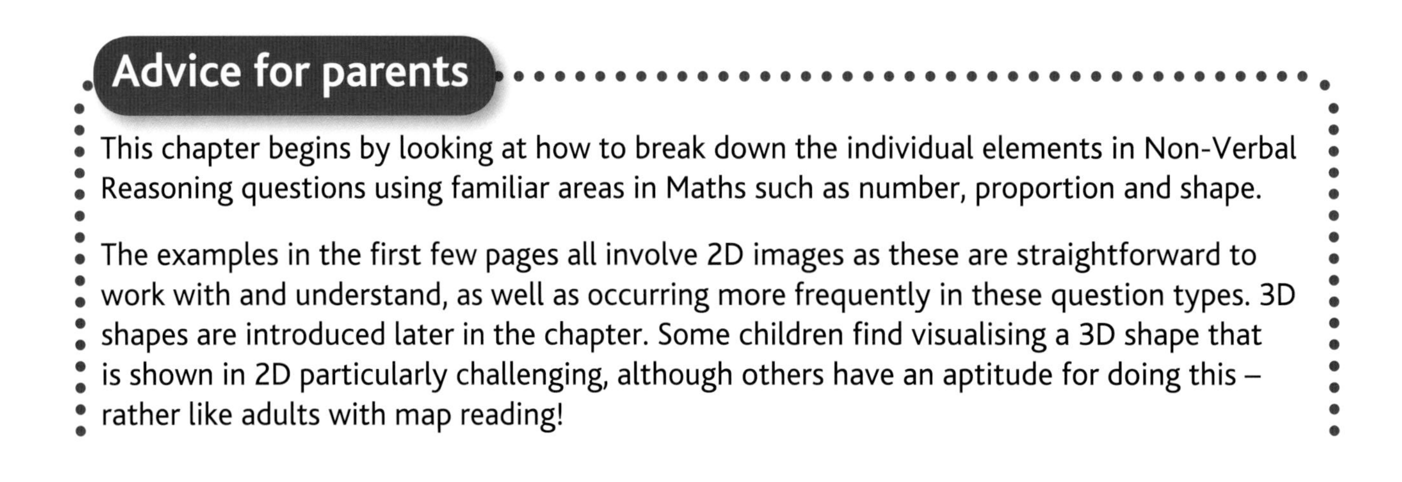

Advice for parents

This chapter begins by looking at how to break down the individual elements in Non-Verbal Reasoning questions using familiar areas in Maths such as number, proportion and shape.

The examples in the first few pages all involve 2D images as these are straightforward to work with and understand, as well as occurring more frequently in these question types. 3D shapes are introduced later in the chapter. Some children find visualising a 3D shape that is shown in 2D particularly challenging, although others have an aptitude for doing this – rather like adults with map reading!

If your child struggles with this skill, try looking at groups of everyday objects together. If you have any building blocks, try stacking them and drawing 2D views – stock cubes and sugar cubes work just as well. Some children find that physically picking up shapes and looking at them from different angles, as well as cutting out nets to form shapes, is helpful too.

Thinking skills and games

The following games can help to develop the skills featured in this chapter.

Dice nets

Trace this net of a cube and transfer it to a piece of thin card to create your own dice. Without folding it up, see if you can add dots to represent the numbers. *Remember, opposite sides of the dice should always add up to 7* (see answers).

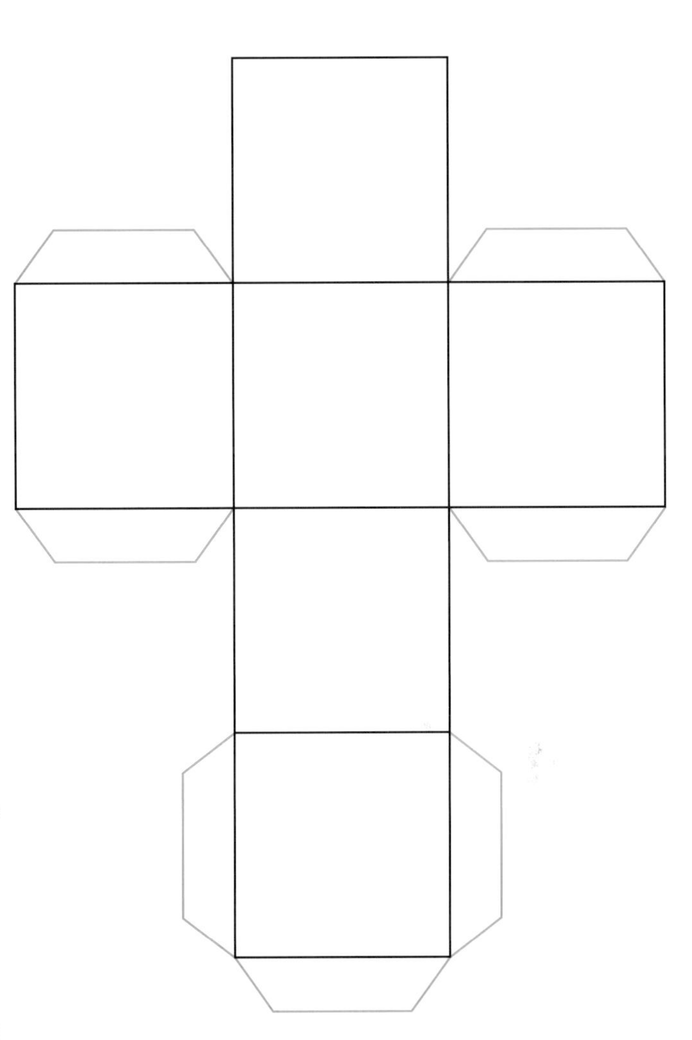

Picture this!

Look for websites where you can print out different nets. Many of these allow you to put your own images or photographs on the faces.

Find two different nets for a cube (one like the 'dice' above and another variation), a square-based pyramid, a tetrahedron and a prism. Now either print the blank nets and stick photos on each face or place photos electronically on the faces before you print them to follow a theme. The themes could be friends, pets, famous people or holidays.

Practise folding and unfolding them to become familiar with how the faces go together. You may find that some of the pictures are upside down when you fold them up so experiment with turning the pictures around on the nets until all the pictures are the right way up.

Sweet memory

Use a pack of sweets that have a variety of shapes or colours. Put all the sweets in a bowl and take it in turns to put ten sweets onto a clean surface.

Give your friend or parent ten seconds to memorise the sweets. Ask them to turn away and, when they are not looking, take one sweet away and hide it. If they can identify the missing sweet, they are allowed to eat it. If they cannot, then it's all yours!

Working with numbers

Skill definition: Find number patterns hidden in different pictures.

- Many Non-Verbal Reasoning questions use number patterns. You may see number patterns created in any of the following ways:
  - sides or angles in a shape
  - lines or intersections of a line
  - shapes, both 2D and 3D
  - gaps in a picture (an easy one to miss so look out for it!)
- Although many number patterns just involve matching numbers, the more complex questions include addition and subtraction, as in the example described below. These patterns feature in many of the questions you will see in this book.

Remember: There are many possible numbers relating to shape. If you cannot spot a pattern, consider the number of common shapes, different shapes, shaded shapes, sides and vertices.

Method

Look at the first three pictures and decide what they have in common. Then select the option from the five on the right that belongs to the same set. Circle the letter beneath the correct answer.

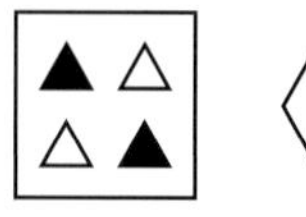
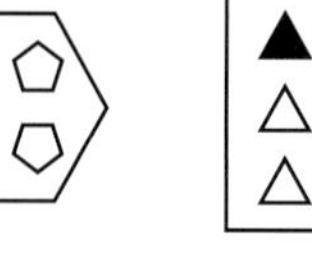
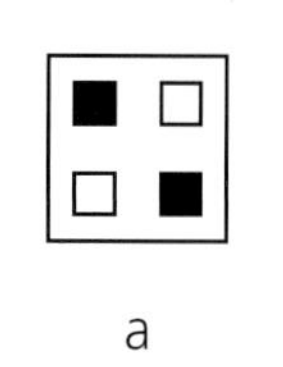
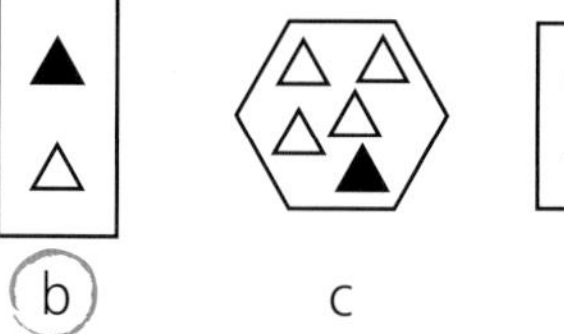
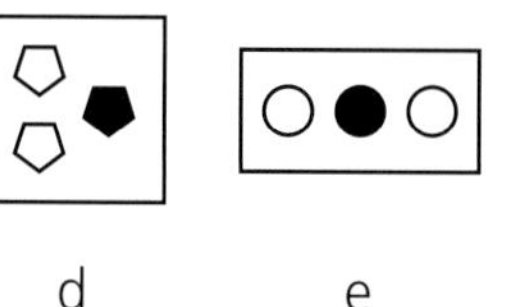

a (b) c d e

- Begin by looking at the first three pictures to see if you can find any common features.
  - All three pictures have smaller shapes inside a larger shape, but the shapes are different.
  - Some of the smaller shapes are shaded, but the number varies.
- Now look at the answer choices. All the answer choices match these common features, so this means that there is another feature that you still need to find.
- The next step is to look for **connections between features**. Think about whether:
  - the number of shaded shapes match anything else
  - the number of unshaded shapes match anything else
  - the number of sides or vertices in the small shapes match those in the larger ones.
- The number of shapes (shaded or unshaded) seems to follow a pattern. The answer is the **number of sides** of each small shape is **one less** than the number of sides in the large shape in each of the first three pictures. This makes **b** the correct answer because the outer rectangle is a four-sided shape and the small triangles are three-sided shapes.

This is a difficult question because of the number of **distractors** (that is, changing features that are not part of the answer).

The only way to solve these tricky problems is to look at all the options one at a time without getting put off. There will be an answer – it might just need a bit of detective work to find it.

Train

1 What is the number connection between these pairs of pictures?

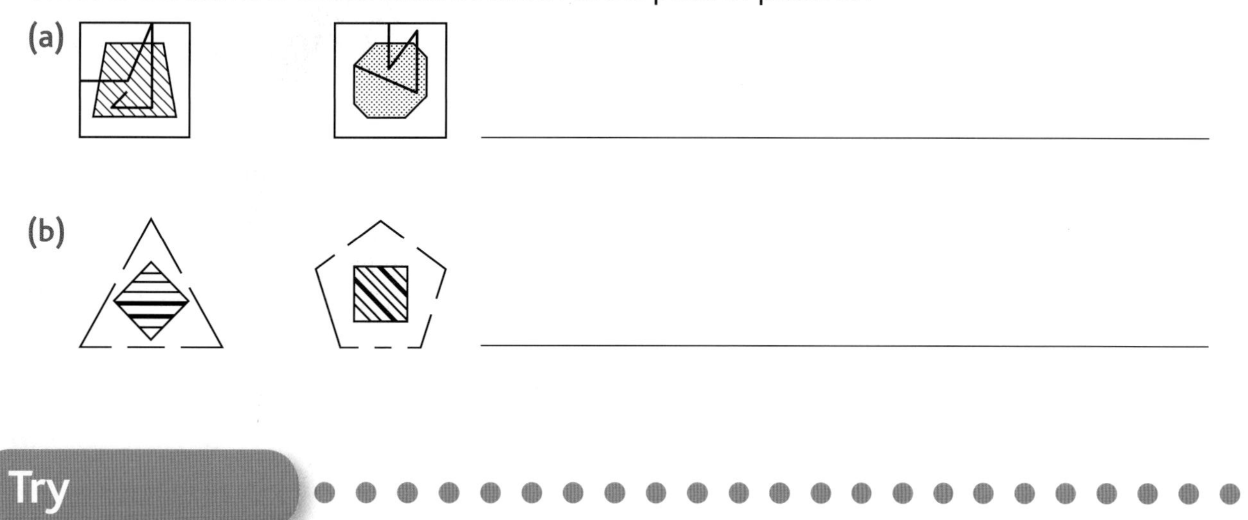

Try

Look at the first two pictures and decide what they have in common. Then select the option from the five on the right that belongs to the same set. Circle the letter beneath the correct answer.

Test

Test time: 02:00

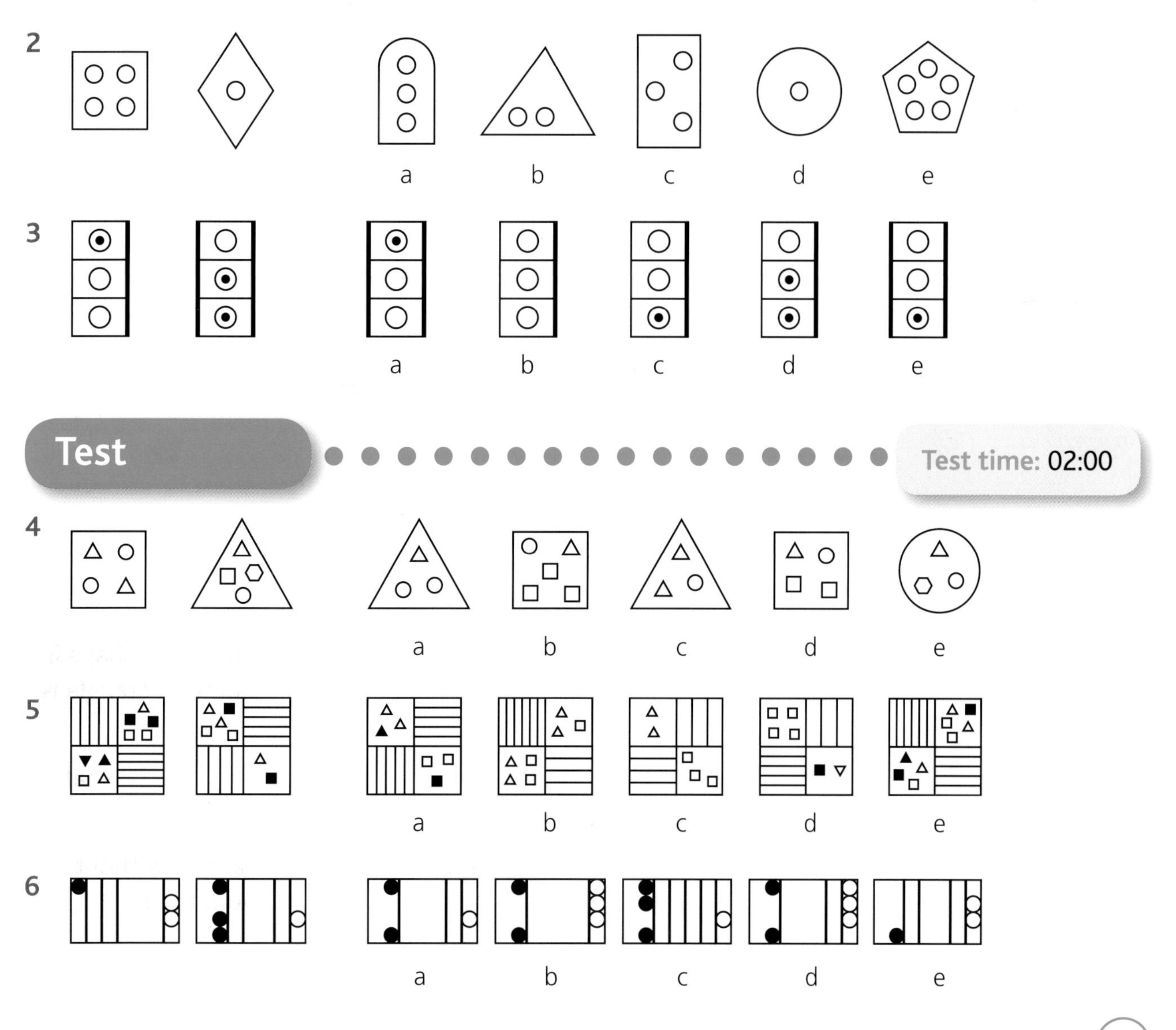

Problems involving scale

Skill definition: Find patterns with varying lengths, sizes and proportions in different pictures.

- Patterns that use different sized shapes and lengths of lines are common and are often hidden by a number of distractors so you need to be extra vigilant.
- Patterns may be created in any of the following ways:
  - different sizes of shapes (often concentric)
  - varying line lengths
  - fractions of a shape, including those that overlap (also see page 48)
  - groups of shapes in various proportions (which can also be seen as number problems).
  - The examples shown here are all *Applying changes* questions, although the skills used feature in many other questions you will see in this book.

Questions involving length can be tricky because they often include other elements. The difference between these two pictures is the length of the line at the base.

Method

Look at the two pictures on the left connected by an arrow. Decide how the first picture has been changed to create the second. Now apply the same rule to the third picture and circle the letter beneath the correct answer.

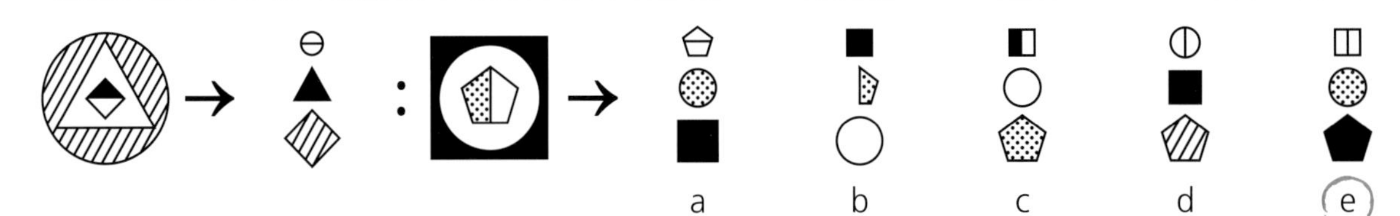

- This question uses shading to trick you into missing what is really important. This question is actually about the size and proportion of the shapes.
- Look at the first picture, ignoring the shading.
  - There is a large circle, a smaller triangle and an even smaller diamond.
  - Notice that the diamond is divided into two halves with one-half shaded.
- The shading on the diamond makes it look like the triangle in the second picture but, as all three shapes are also in the second picture, this shading is a distraction.
- Break down the changes between the first and second pictures.
  - The largest shape (the circle) becomes the smallest shape at the top, gaining the division line from the middle shape.
  - The medium-sized shape (the triangle) stays in the middle.
  - The smallest shape (the diamond) becomes the largest shape at the bottom and loses the division line.

- Now you can look at the pattern of shading.
  - The shading from the largest shape moves to the shape at the bottom.
  - The shading from the medium-sized shape moves to the shape at the top.
  - The shading from the smallest shape moves to the middle shape.
- Once you apply these rules carefully to the picture after the colon, you can see that the correct answer option is **e**.

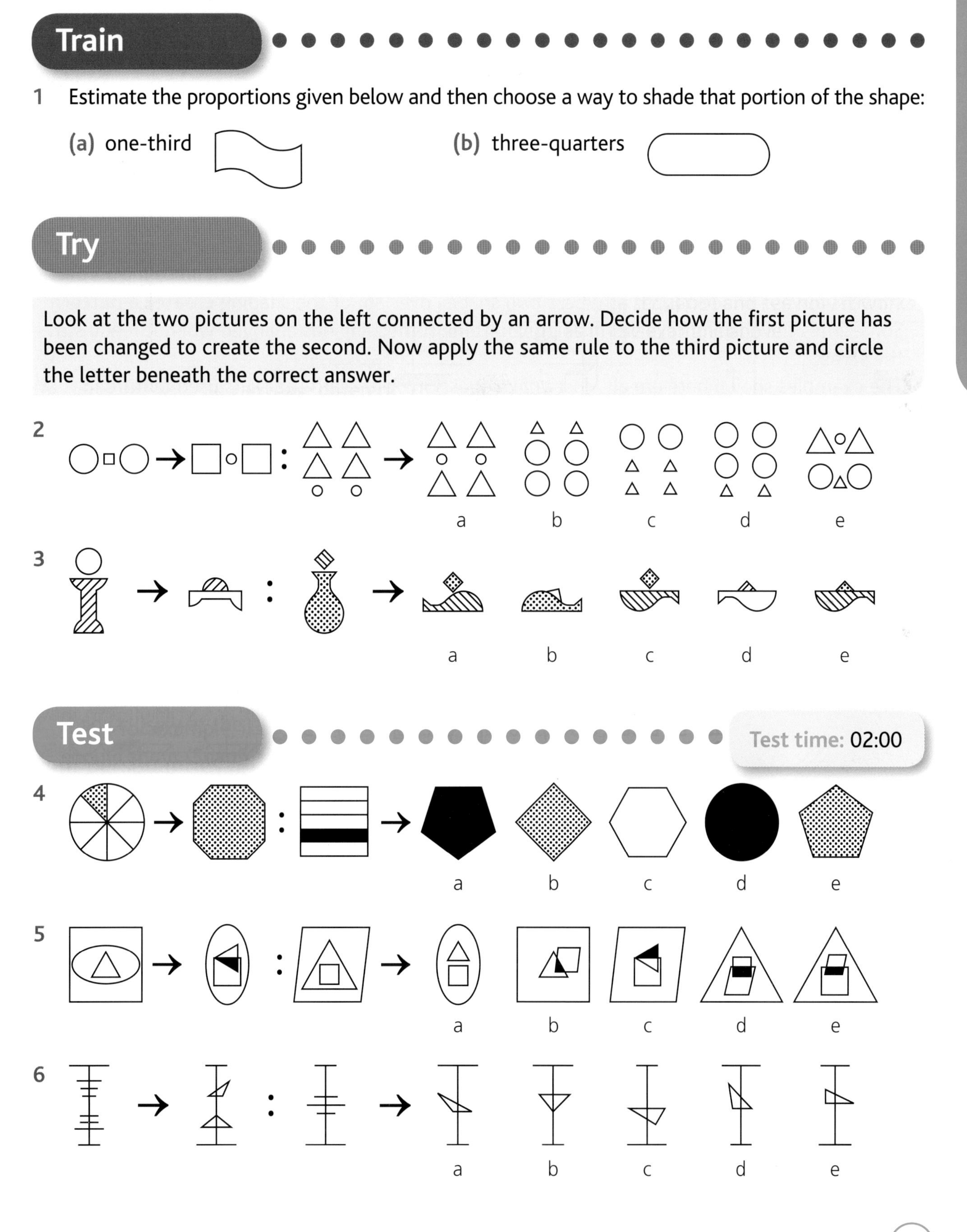

Train

1 Estimate the proportions given below and then choose a way to shade that portion of the shape:

(a) one-third

(b) three-quarters

Try

Look at the two pictures on the left connected by an arrow. Decide how the first picture has been changed to create the second. Now apply the same rule to the third picture and circle the letter beneath the correct answer.

2

a b c d e

3

a b c d e

Test

Test time: 02:00

4

a b c d e

5

a b c d e

6

a b c d e

Shapes and shading

Skill definition: Find patterns made by shapes and the shading within them.

- You have already seen some patterns using shapes. Now you are going find out how the **properties of shapes** can also be important. You may see patterns created by shapes in any of the following ways:
  - regular and irregular shapes
  - 2D and 3D shapes
  - single shapes with different properties, such as triangles with different angles
  - shapes with parallel, straight, concave and convex sides
  - symmetrical shapes (these are covered more fully in *Symmetry and vertical reflections* on page 50)
  - shapes that connect together and overlap to create new shapes (this also relates to *Translations*; see pages 58–62).

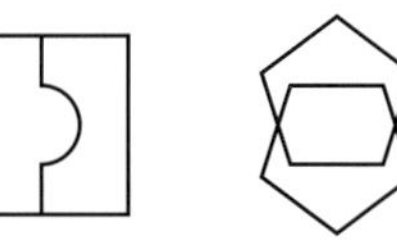

- Shading is a common element in shape questions. Look for patterns where a certain area of the shape is **always shaded** (such as where two shapes overlap) or the shading **creates a pattern**. Sometimes the shading **reverses** making you think something has changed when it really stays the same.
- The examples shown here are all *Most unlike* questions, although the skills used feature in many other questions you will see in this book.

The direction of any striped shading is often important in finding the correctanswer.

Method

Look at these pictures. Identify the one that is most unlike the others. Circle the letter beneath the correct answer.

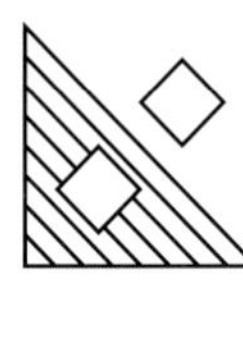

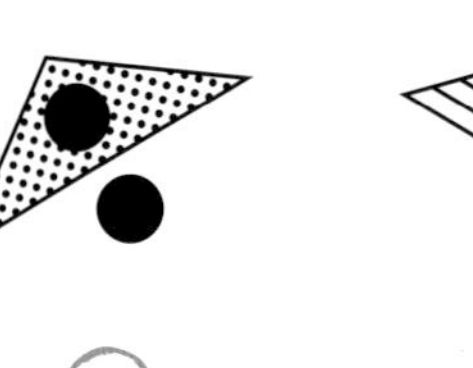

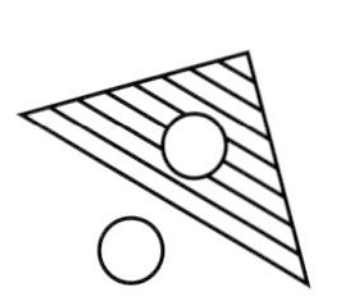

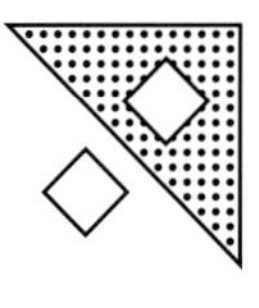

a b (c) d e

- Look at the small circles and squares.
  - There are three pairs of squares and two pairs of circles.
  - Two pairs are shaded black and three pairs are shaded white.
  - One shape is inside the triangle on the long edge and one is outside.

  Because no one element is different here, these shapes cannot hold the clue to the answer.
- The relationship between the shading in the triangles and smaller shapes is worth looking at as there may be a connection between the patterns. However, the black and white small shapes appear with both the striped and the dotted shading of the larger shapes, so this is not a clue either.
- The answer actually relates to the triangles. All the triangles are right-angled triangles apart from option **c** which is an isosceles triangle. Sometimes the answer is quite simple – you just need to be able to see past the distractors.

Train

1 (a) What do the two stars have in common? ______________________________

____

(b) Draw another star that follows the same rule.

Try

Look at these pictures. Identify the one that is most unlike the others. Circle the letter beneath the correct answer.

2

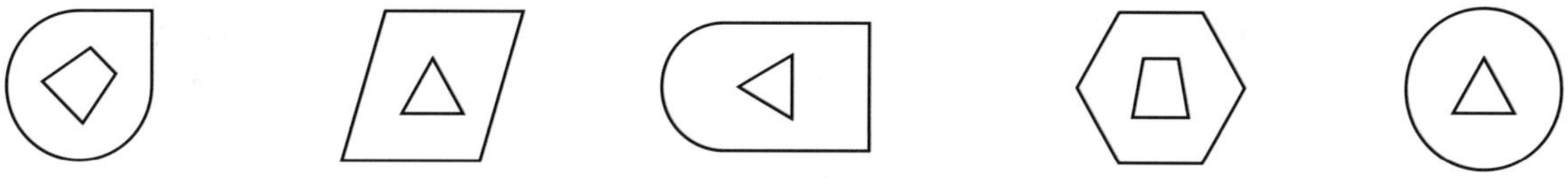

a b c d e

3

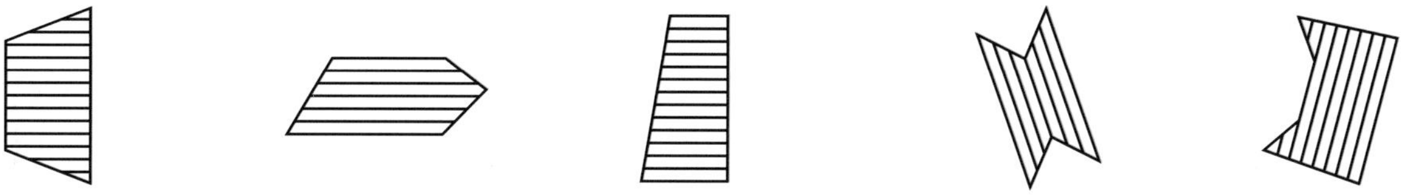

a b c d e

Test

Test time: 02:00

4

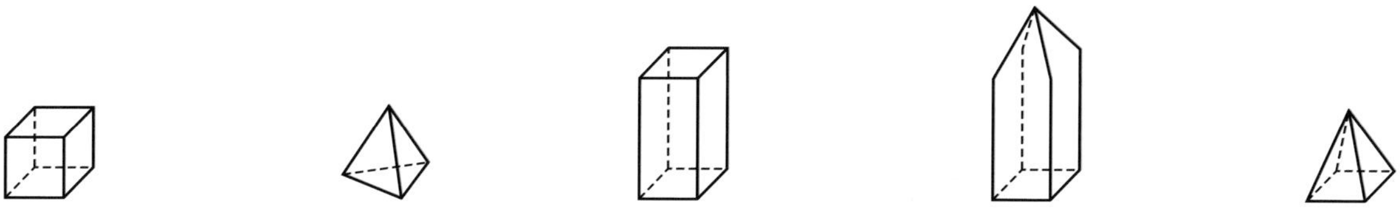

a b c d e

5

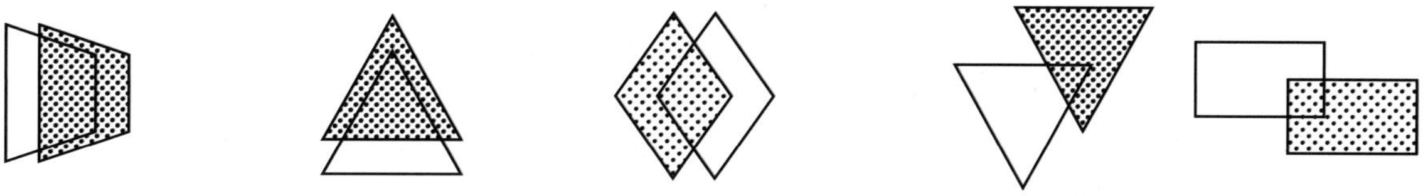

a b c d e

6

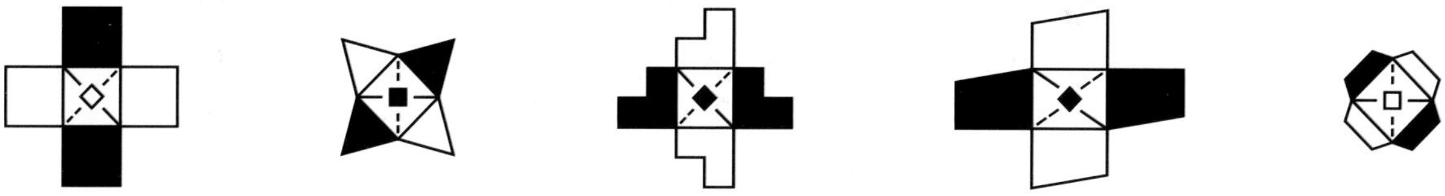

a b c d e

Line styles and angles

Skill definition: Find patterns made by different styles of lines and the angles between them.

- As you have already seen, lines can be used in Non-Verbal Reasoning to represent numbers, show proportions and construct a variety of shapes. Lines can also have **different styles** and can be used to **create patterns**. Lines can be shown in a variety of styles.
  - The line thickness can change or there may be more than one line (often on the tails of arrows).
  - Lines can be solid, dashed or dotted.
  - The shape of the line may change to become curved or jagged.
- The **angles between** lines can also be important. Angle is a common element in questions where there are patterns made with lines: for example, a common theme in a set of pictures may be the angle where **two lines meet**. Angles also occur where lines **intersect**, so look closely at these too.
- The examples shown here are all *Matching features* questions, although the skills used feature in many other questions you will see in this book.

The number of curves or jagged shapes on a line is often a feature in questions where number is the clue to the answer. Here three jagged shapes on a line equal the number of circles.

Method

Look at the first three pictures and decide what they have in common. Then select the option from the five on the right that belongs to the same set. Circle the letter beneath the correct answer.

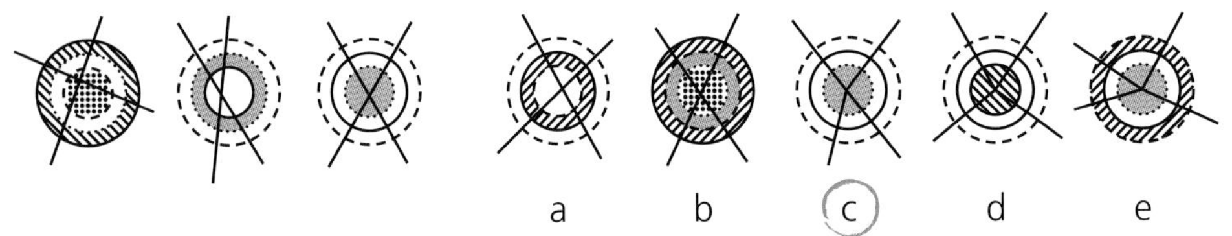

- This question has three examples but it works in exactly the same way as the version with only two. In this question, there are connections of both **line style** and **angle**.
- Look at the line style of the three concentric circles in the first three pictures.
  - Each has one dotted, one dashed line and one solid line.
  - So we know that there will be one dotted, one dashed and one solid line in the answer. This means the possible answers are **b**, **c** and **e**.
- Now look at the angles created by the two intersecting lines. This angle changes, so cannot be common feature. However, because the lines are straight, the opposite angles are equal. Looking at the answer options, you will see that some of the opposite angles are not equal because the intersecting lines are not all straight. So the only answer option that could be correct is **b**.
- This means that the shading is a distractor.

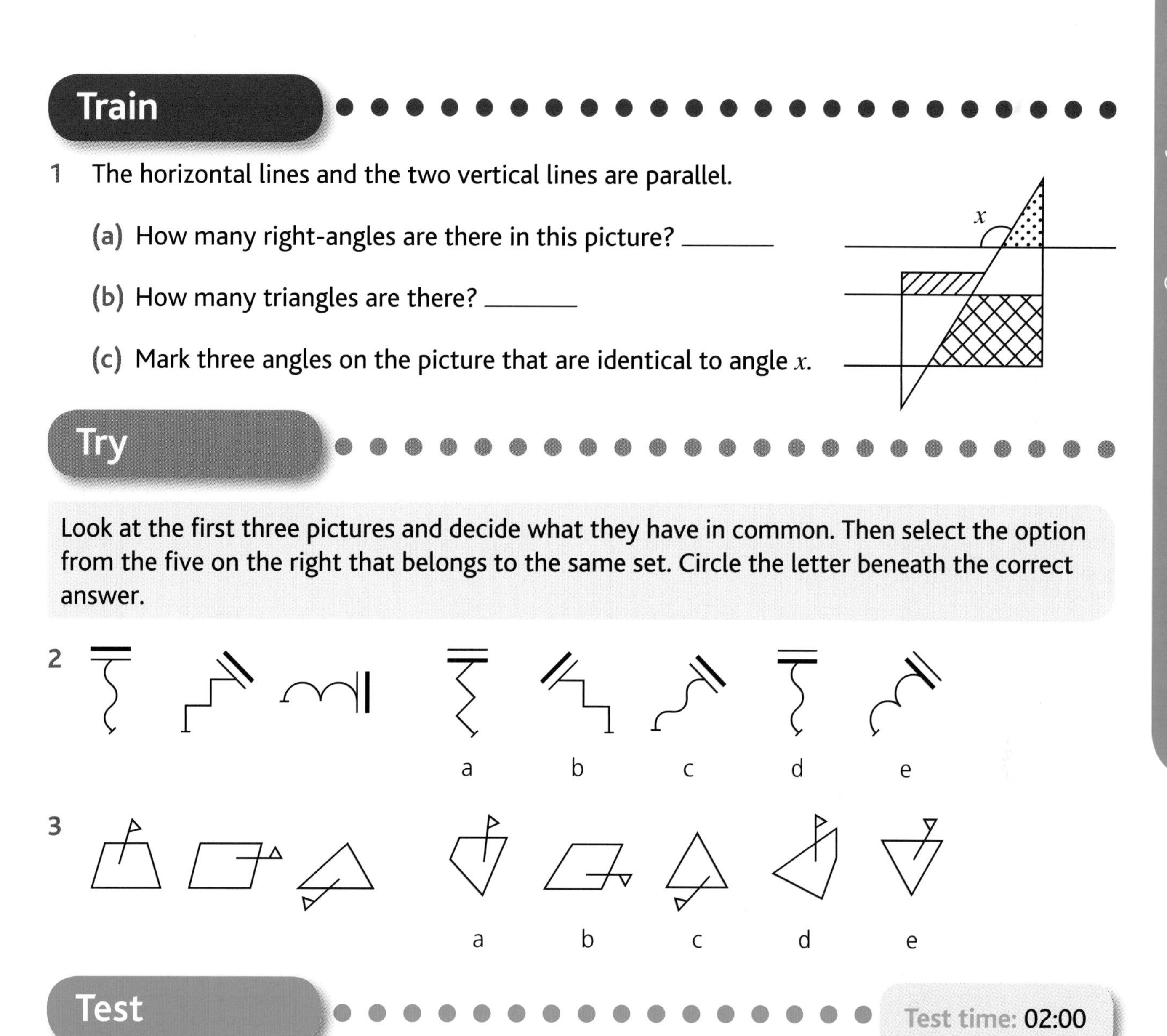

Train

1 The horizontal lines and the two vertical lines are parallel.

(a) How many right-angles are there in this picture? ______

(b) How many triangles are there? ______

(c) Mark three angles on the picture that are identical to angle x.

Try

Look at the first three pictures and decide what they have in common. Then select the option from the five on the right that belongs to the same set. Circle the letter beneath the correct answer.

2 a b c d e

3 a b c d e

Test

Test time: 02:00

4 a b c d e

5 a b c d e

6 a b c d e

3D views of 2D pictures

Skill definition: Match a 3D shape to the net provided.

- All these questions ask you to imagine what a net would look like when it is folded up into a 3D shape.
- Unlike the other questions in this chapter, the shapes do not change from picture to picture. You are just being tested on your understanding of how 3D shapes are put together.
- Nearly all nets questions use the standard net of a cube (see below), although a few questions show variations. Sometimes tetrahedron nets are used (see right) – these are simpler to work out as they have fewer sides.

Working out the folds in nets

Imagining how to fold a net into a 3D shape can be tricky, so cutting out printed nets and folding them up can be really helpful.

Standard nets

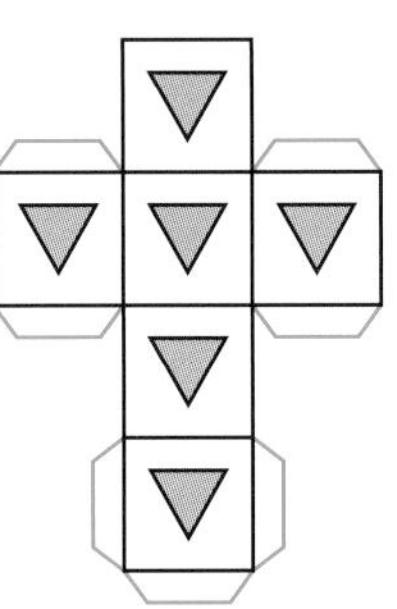

First trace and cut out the net of a cube on page 27.
You can cut off the blue tabs if you find this easier to work with.

- Now work with your net to follow these instructions.
  - Draw triangles onto the net, copying the diagram shown here. These triangles act like arrows to show how the **patterns** relate to each other when the net is folded.
  - Slightly fold down the sides that are on the right and left. You will see that they form three faces that run around three sides of the cube. The arrows all continue to point downwards. **When two faces are joined side by side, the direction of the pattern cannot change**.

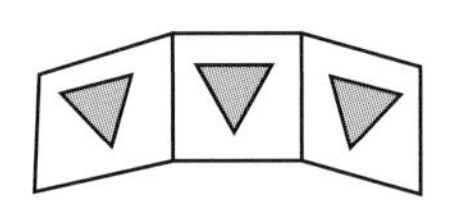

  - Next, roll up the net into a cylinder to see how the direction of the arrows works. You will see that they form a **continuous loop**.

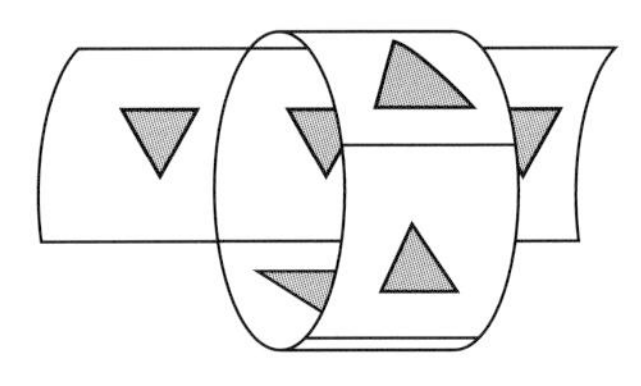

- Unfold the net and copy it onto a piece of paper. Add the following features.
  - Draw two eyes on the net so that it looks like an elephant with the ears pointing out, hair pointing down and a pattern down the trunk.
  - Next add two sets of arrows: draw a blue pair from the ears upwards to form the top of the elephant's head and draw a red pair downwards to form cheeks. These blue and red arrows **show how the sides link together**.
  - Finally draw two green arrows from the outside of the ears to the bottom of the trunk. The elephant looks very fat now! These green arrows **show how the sides link to the base**. Now imagine folding the elephant's trunk under and round to touch the back of his head to form the net.

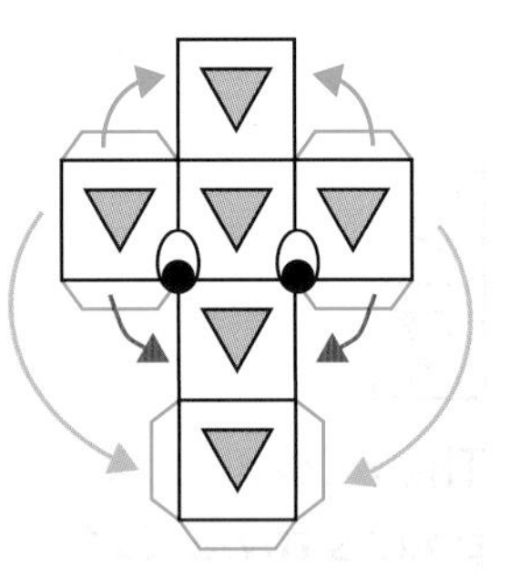

Explore how the arrows work by pinching adjacent sides together. Putting numbers on the faces and different coloured dots on touching sides can also be helpful.

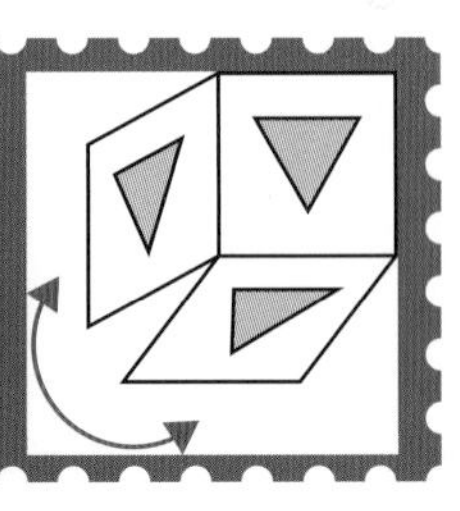

Nets with four faces in a row

There are five possible nets of a cube that have **four faces in a row**. When you spot these different nets, think elephant! Practise these questions by drawing the nets and adding the triangles as you did before to show the direction of the pattern.

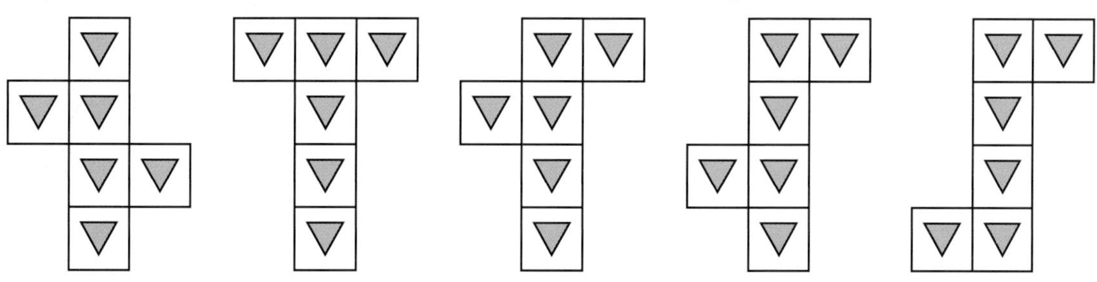

- The same rules work for **faces next to each other** (remember the red and blue arrows for the elephant's head and cheeks), and for **the outer faces** (the green arrows). The ears might have moved about a bit, but you can still draw in the coloured arrows quite easily!

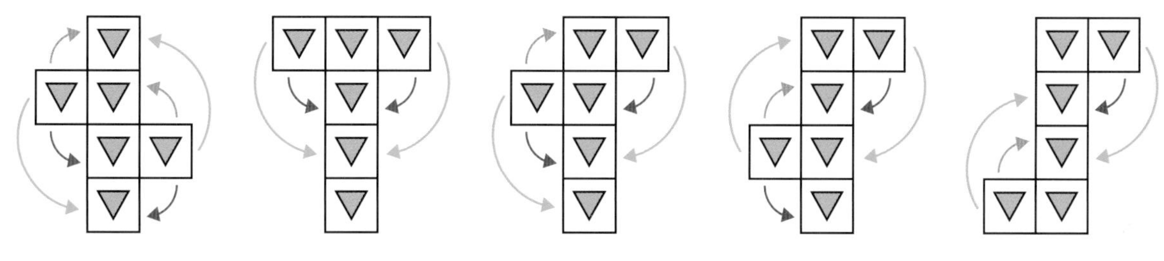

- There are still a **few sides** in the final four nets that are **not connected** by arrows. The **different coloured circles** show the **sides that join together**. Put dots on these edges so that you can see what happens when you fold the nets.

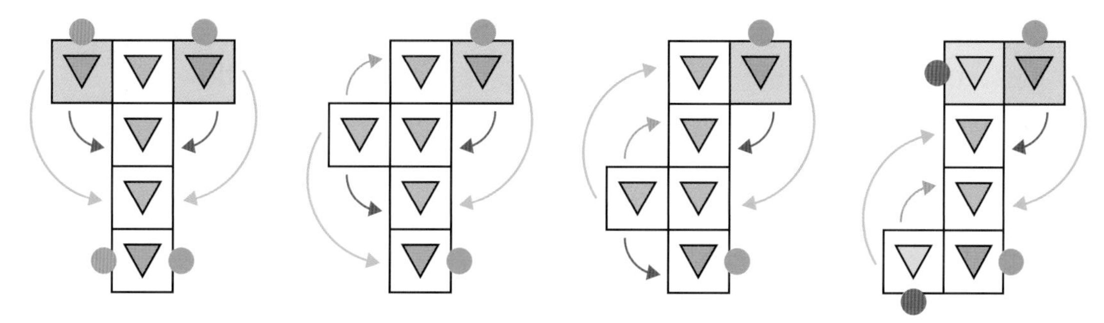

- The coloured triangles show which folds follow the same rules. The sides with blue and orange circles fold with the bottom triangle pointing inwards across the top of the grey faces.

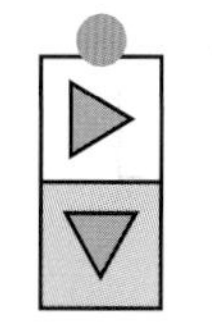

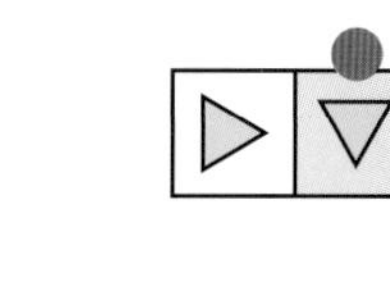

- The **odd-one-out** is the **last net**. The **bottom green triangle points towards the top blue face**. This is easy to remember because the net looks like a **snake** with a bright red tongue and green tail!

Nets with two or three faces in a row

- There are four possible nets of a cube that have **three faces** in a row and one possible net that has **two faces** in a row. Practise drawing them and adding **triangles** to **follow the patterns**. Although the central faces do not form a loop, you can still work out most of the faces.
- Add red and blue arrows for **faces next to each other** and green arrows for faces that are **two steps away** (look back at how these green arrows work in previous nets).

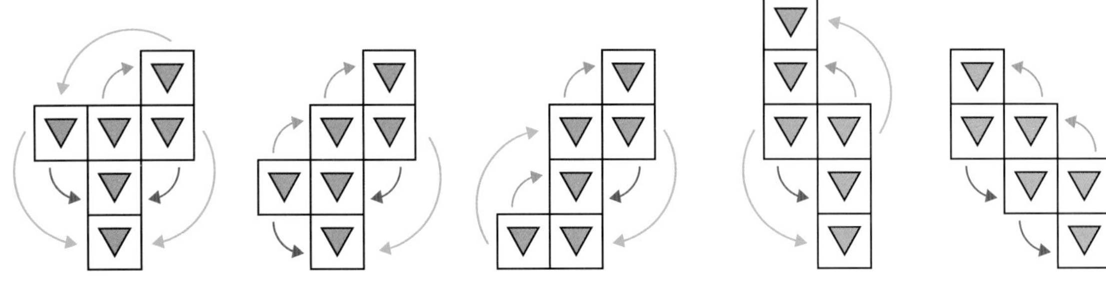

- You can now see the sides that still have to be matched up. The diagram below, together with its key on the right, shows how these sides connect. Look closely at the patterns as you did before. Notice that the last three nets also have triangles that point in **the same direction when folded** (see the faces marked with purple dots).

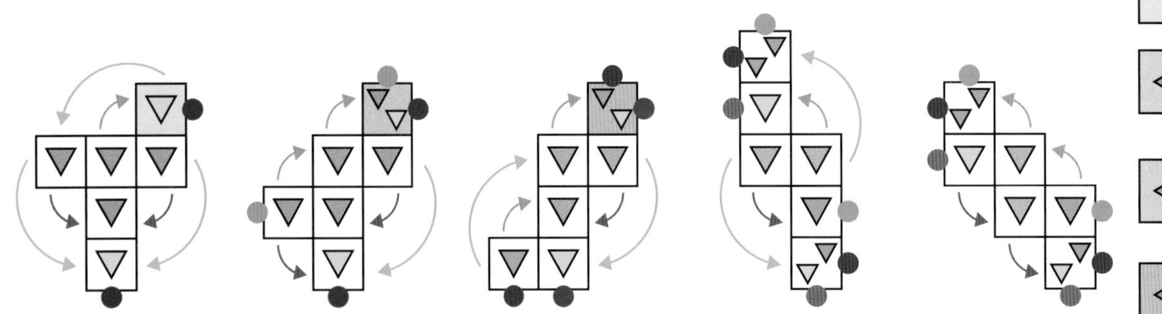

- Practise cutting up, colouring, numbering and folding the nets to understand how they work. The more you play with these shapes and follow the rules described, the easier you will find these questions to answer.

Method

Find the cube that can be made from the net shown on the left. Circle the letter beneath the correct answer.

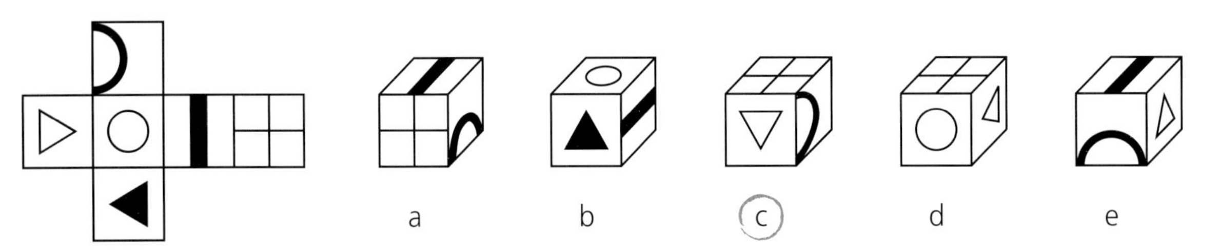

- The quickest way to reach the answer in these questions is by **rejecting** incorrect options.
- Look at the **front face** of each cube together with the **face above it**.
  - Thinking of the **central line** of the net as a circle (as you did with the elephant's trunk on page 36), you can reject options **a** and **d**: the cross does not touch the line in the face above so option **a** is incorrect, and the circle cannot be next to the cross so option **d** is also incorrect.
  - Now look at the **row across** to find more clues. Option **b** cannot be correct because the black triangle does not point at the circle, so this leaves you with options **c** and **e**.

- **Draw arrows** onto the diagram, as in the example on page 36, to help you.
- If you still have problems with seeing how the sides join, imagine pinching the sides and then drawing what you expect to see.

- Option **c** is the only possible answer because the arc in option **e** is in the wrong position when you look at the triangle.

Train

1 Draw three arrows onto this pyramid so that, if the base is the central triangle, you will see all the arrows pointing upwards.

Try

Find the cube from the five options that can be made from the net shown on the left. Circle the letter beneath the correct answer.

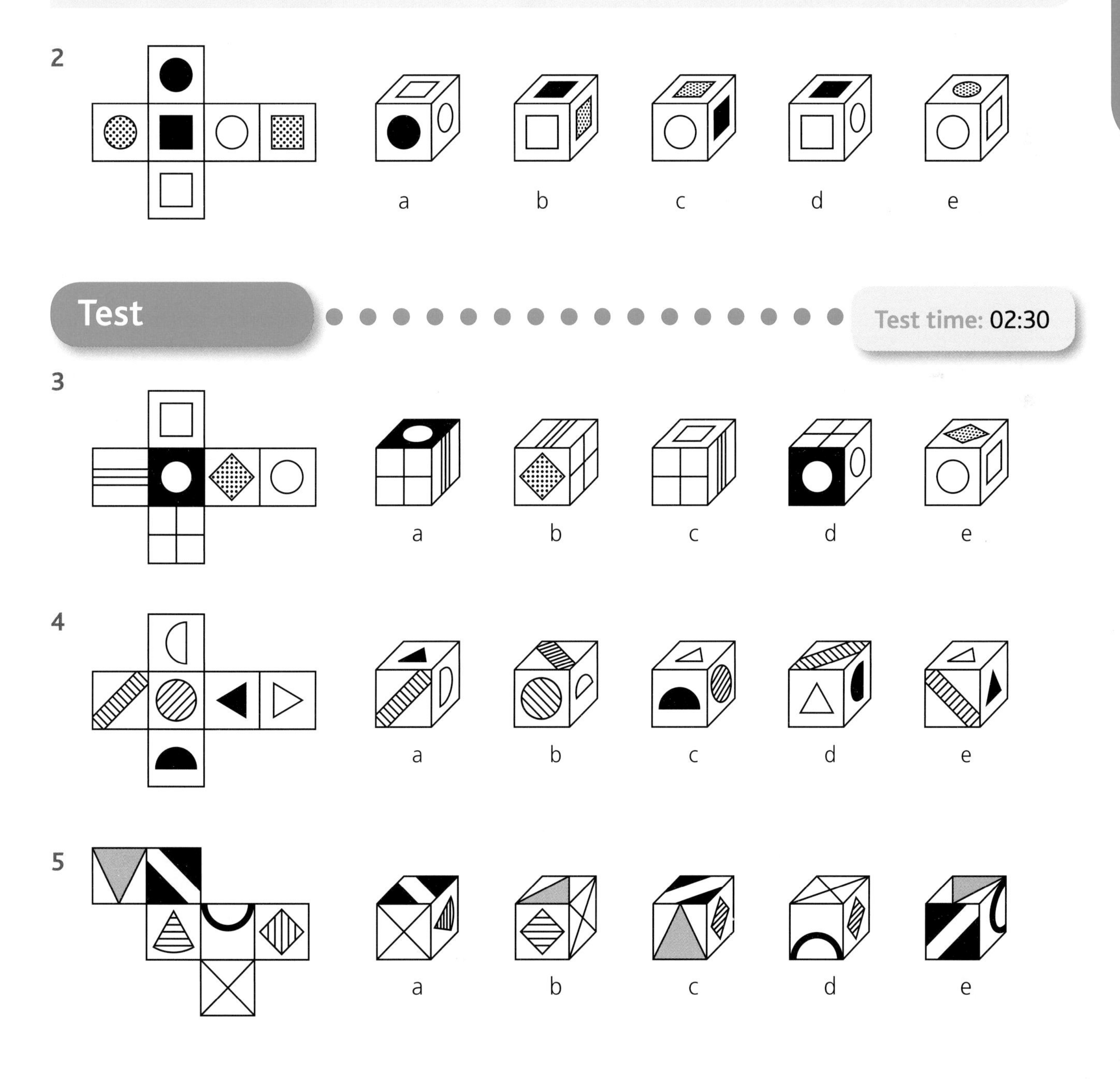

2D views of 3D pictures

Skill definition: Match a 2D net or plan to the 3D shape provided.

- These questions test your ability to imagine what a 3D shape would look like in 2D and test very similar skills to those you have used on pages 36–39. You are likely to come across two types of question.
- The first type is exactly the same as the nets and cube questions, just the other way around. In these problems you are asked to imagine what a **cube** would look like when **opened out to form a net**.
  - Before choosing an answer, always check that the diagram is a true net, as some of these combinations of squares may not fold to make a cube.
- In the second type of question you are asked to imagine what a **set of cubes** will look like when **viewed from above** (that is, a **plan** view) as in the example on the right.

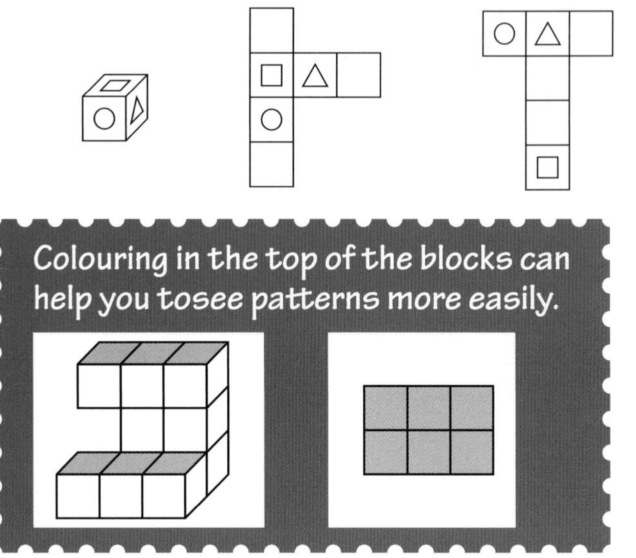

Method

Which of the answer options is a 2D plan of the 3D picture on the left, when viewed from above? Circle the letter beneath the correct 2D plan.

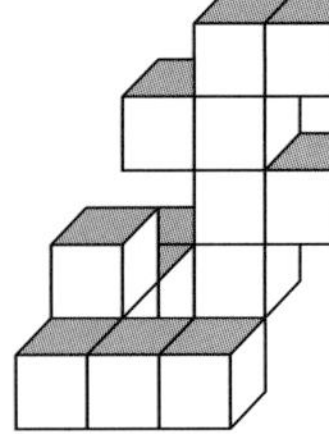

a b c d e

- The key point to remember with these questions is that the height of the 3D picture does not matter as the image is **flattened**. You are looking down on the 3D picture so the only important features are the **number of blocks wide**, the **number of blocks long** and any **gaps**.
- Study the 3D picture in two stages. This will help you to reject the incorrect answer options.
  - First decide how many blocks wide you are looking at, beginning with the front row of blocks. The row is three blocks long so you can immediately reject option **e** which is only two blocks long at the front.
  - Next look at how many blocks the model goes back. Beginning at the right-hand side, it is difficult to guess since the tower of four blocks appears to be **floating** in mid-air. This suggests that there is a **gap** in the shape.
  - Look more closely at the left-hand side. There are definitely two clear blocks you can see in a row, so option **b** must be incorrect. There is then an odd shape, which must be another block (otherwise there would not be a line

Remember: Even if you find it difficult to see the answer, if you have worked through the other options and know they are incorrect, that can be good enough!

drawn in the picture) and a clearer block at the back. So option **c** must be incorrect too since there is a block missing on the left on this plan.

- You are now left with options **a** and **d**. As we can see the blocks on the left-hand side, we know there is a gap in the middle. As the tower of blocks is only one block deep, the answer must be option **a**.

Train

1 These pictures are plan views of everyday items. Write the letter that identifies each item on the lines provided.

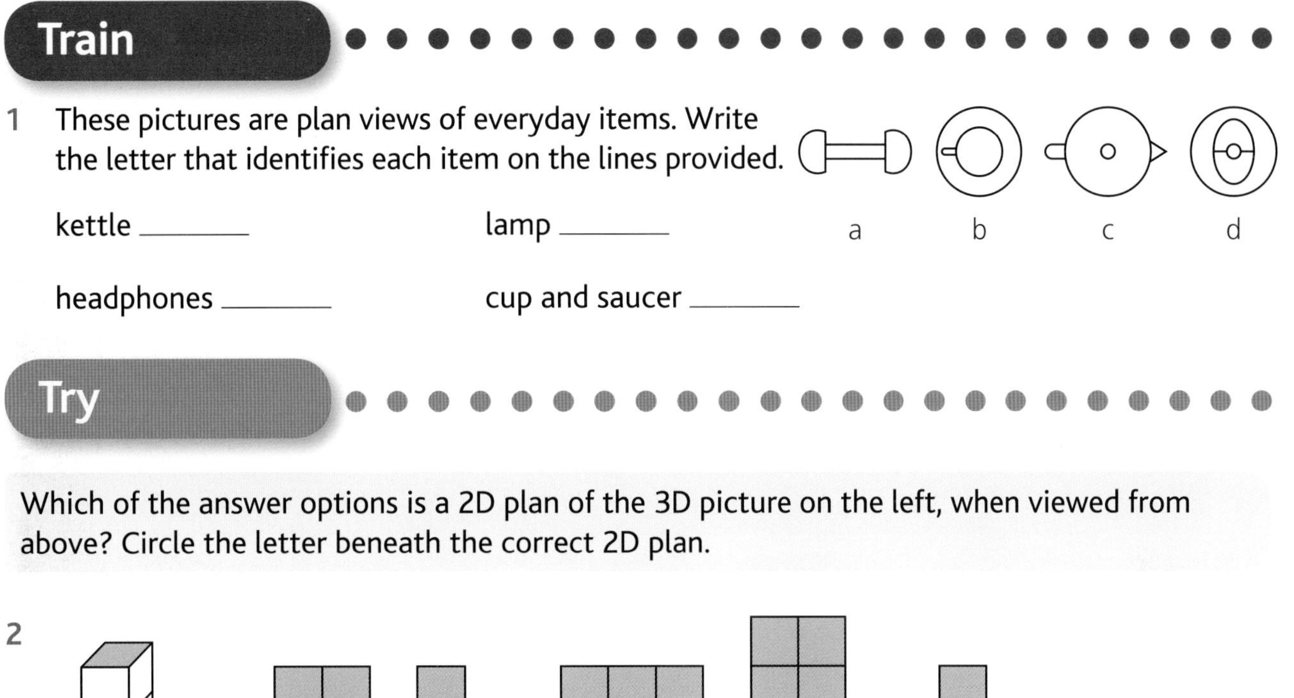

kettle ________ lamp ________

headphones ________ cup and saucer ________

Try

Which of the answer options is a 2D plan of the 3D picture on the left, when viewed from above? Circle the letter beneath the correct 2D plan.

Test

Test time: 02:30

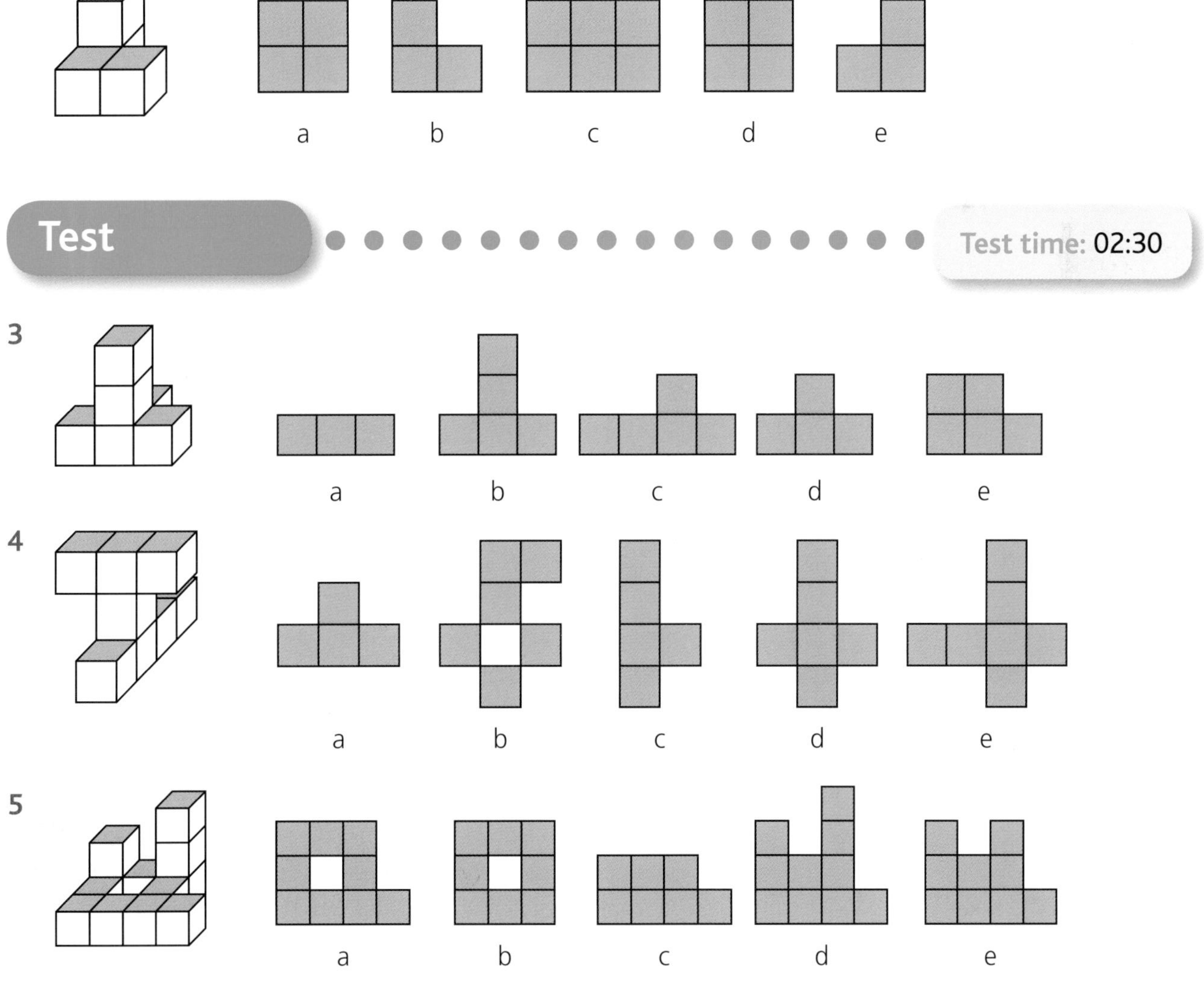

Test 2

Test time: 12:00

Complete this test in the time given above. Each question is worth one mark.

Look at these pictures. Identify the one that is most unlike the others. Circle the letter beneath the correct answer. Example:

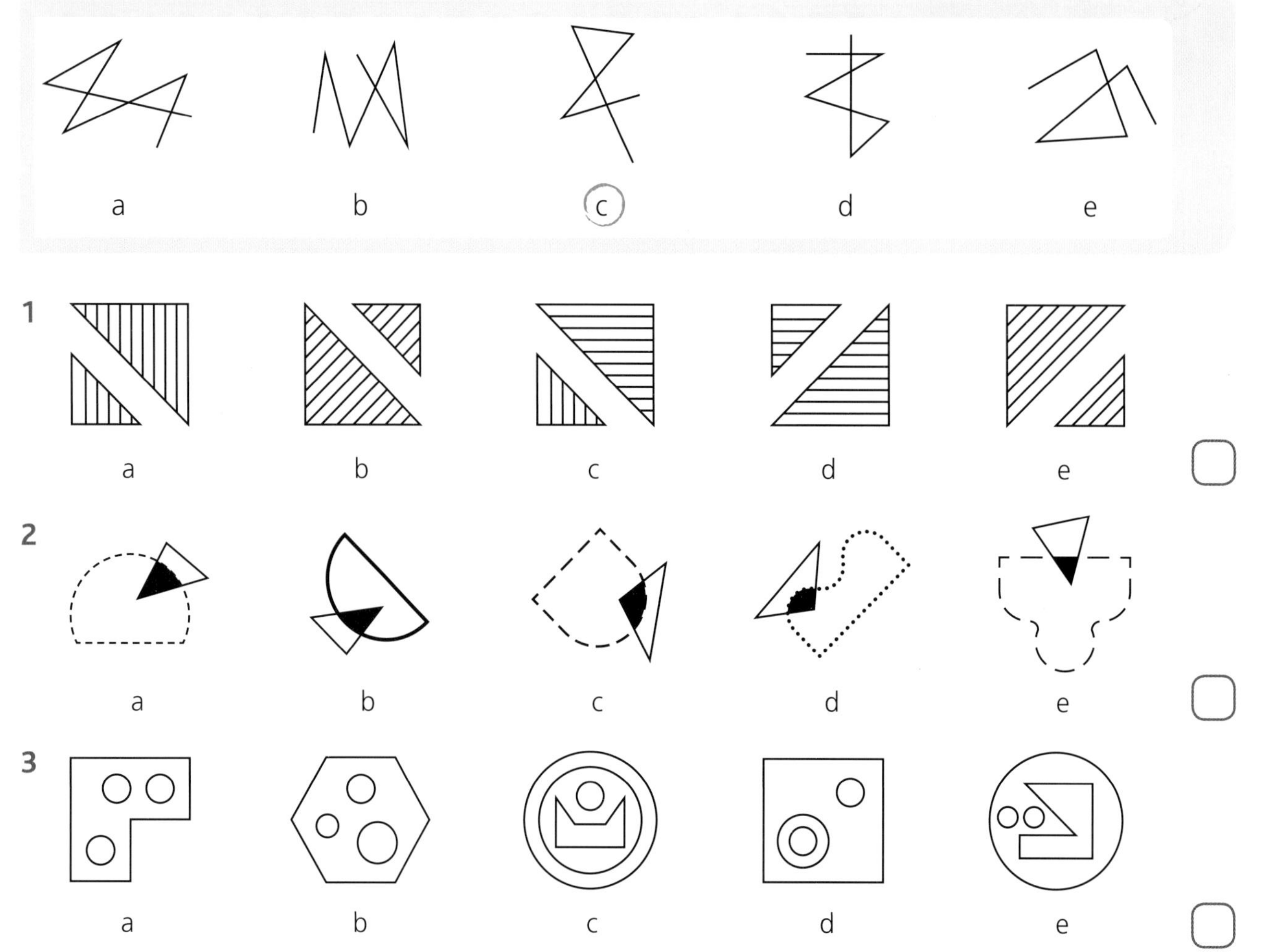

Look at the two pictures on the left connected by an arrow. Decide how the first picture has been changed to create the second. Now apply the same rule to the third picture and circle the letter beneath the correct answer. Example:

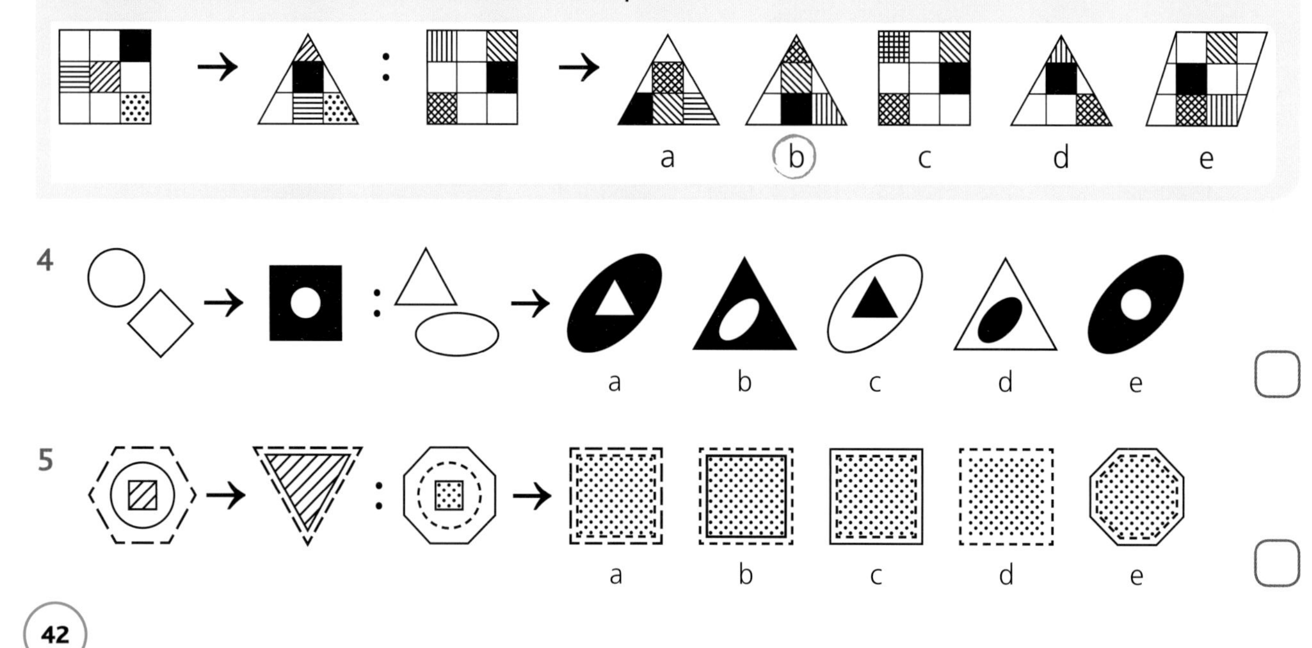

Look at the first three pictures and decide what they have in common. Then select the option from the five on the right that belongs to the same set. Circle the letter beneath the correct answer. Example:

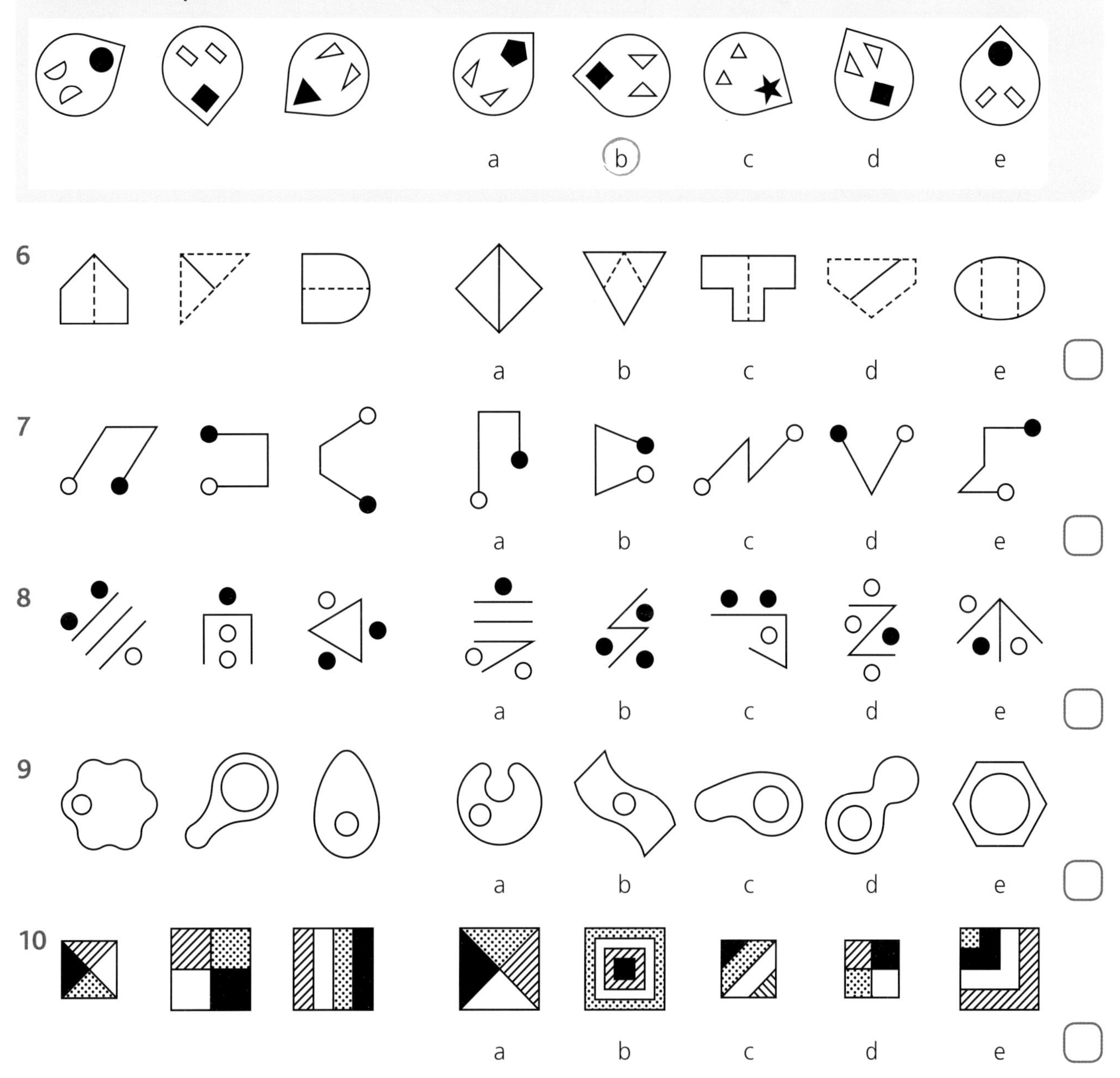

Find the cube that can be made from the net shown on the left. Circle the letter beneath the correct answer. Example:

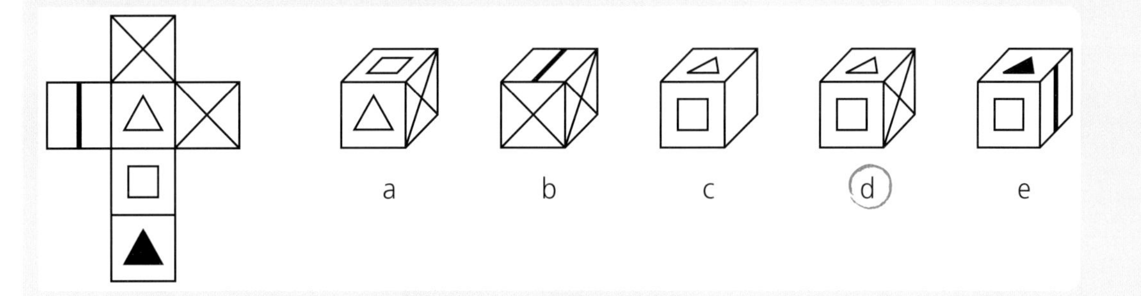

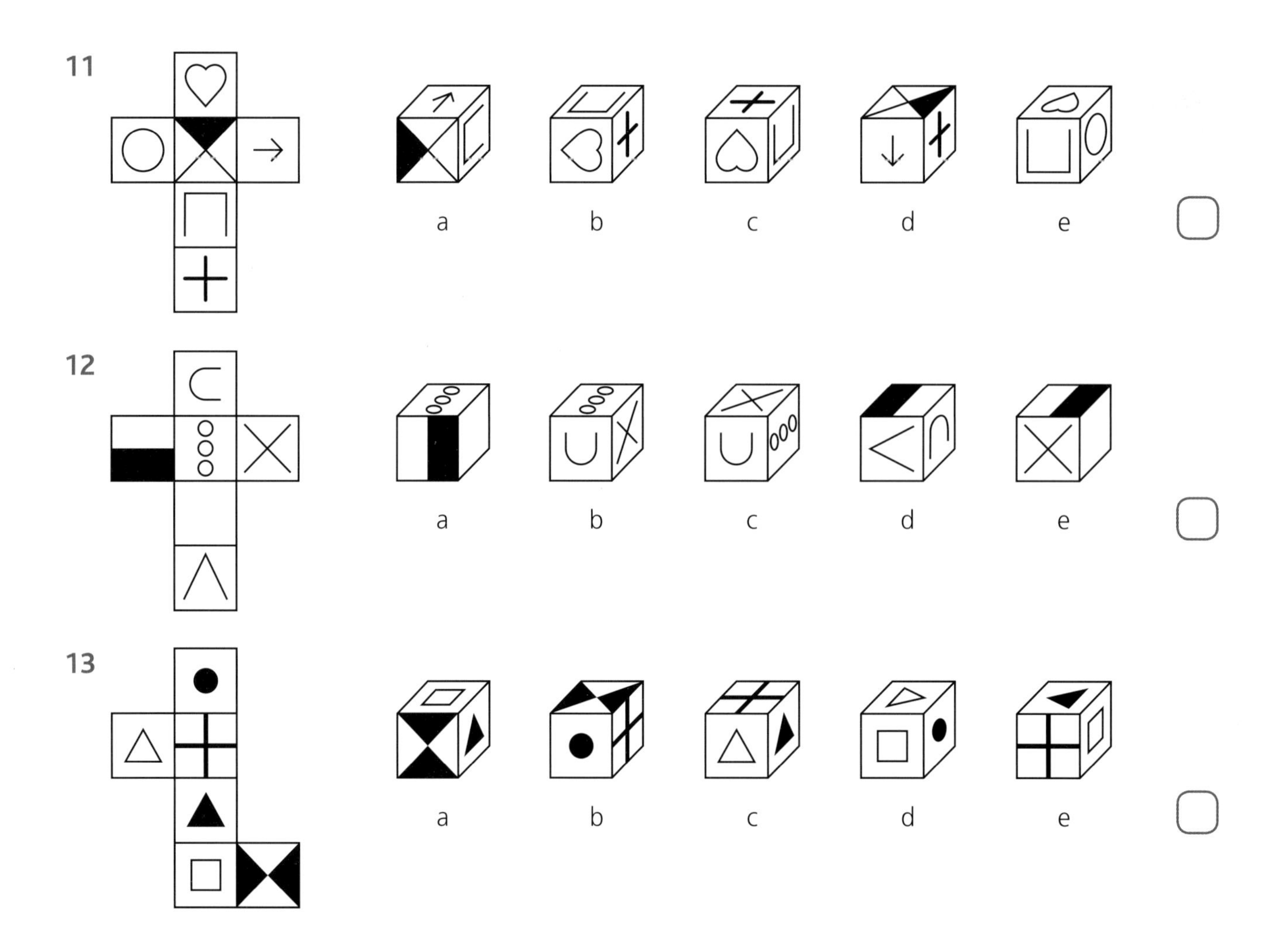

Choose the net that represents the cube when it is unfolded. Circle the letter beneath the correct answer. Example:

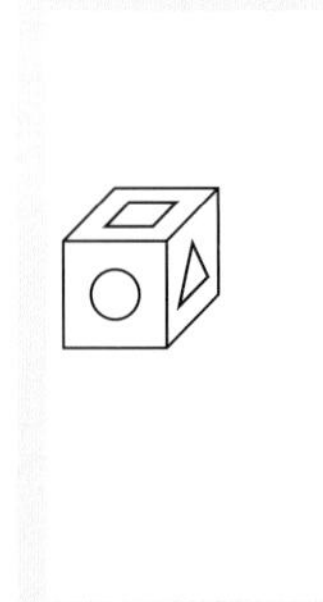
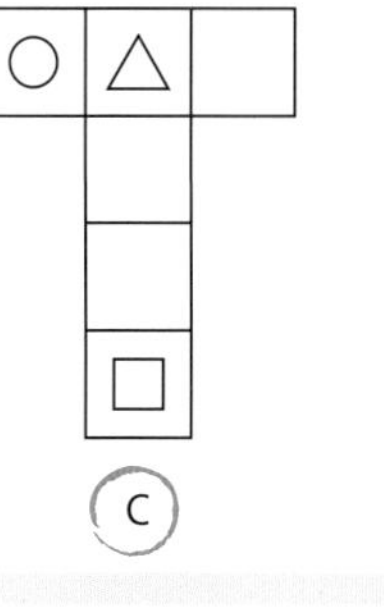
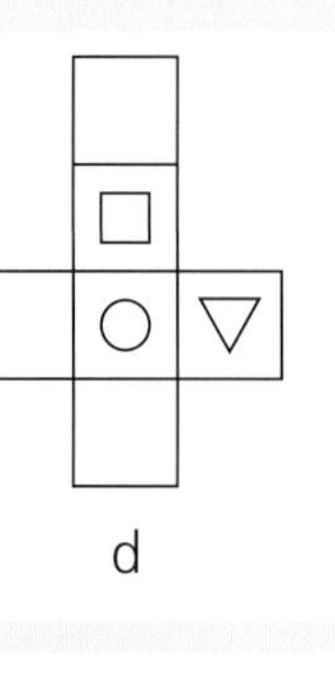
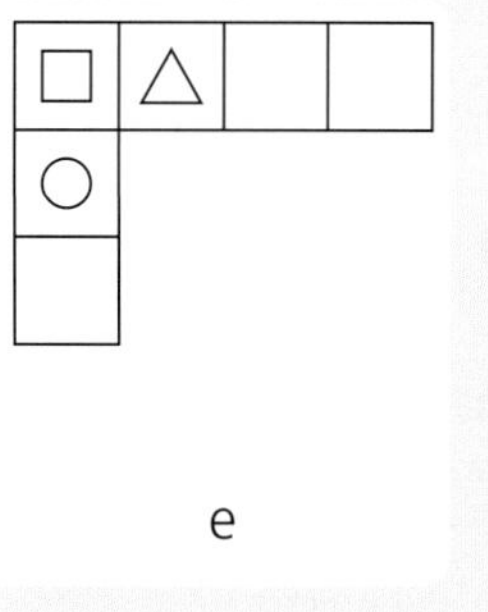

a b c d e

14

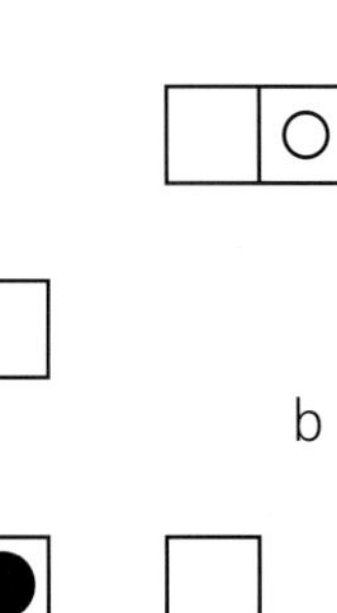
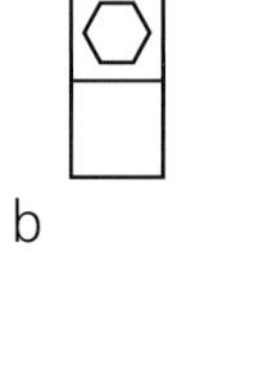
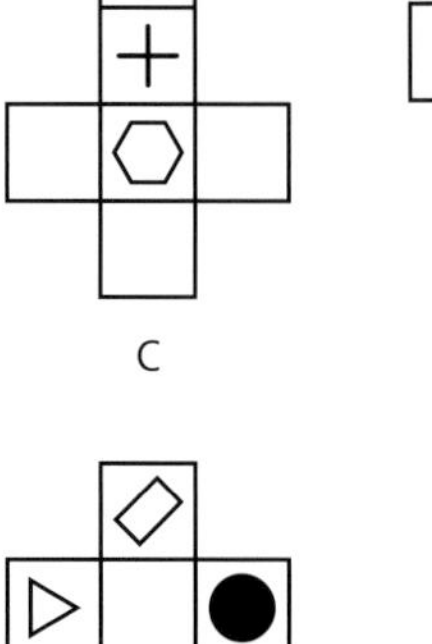
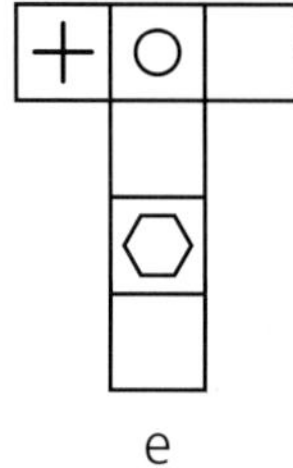

a b c d e

15

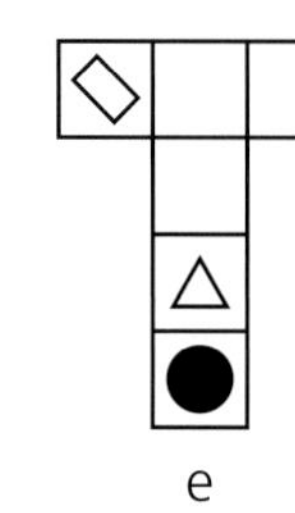

a b c d e

Which of the answer options is a 2D plan of the 3D picture, when viewed from above? Circle the letter beneath the correct 2D plan. Example:

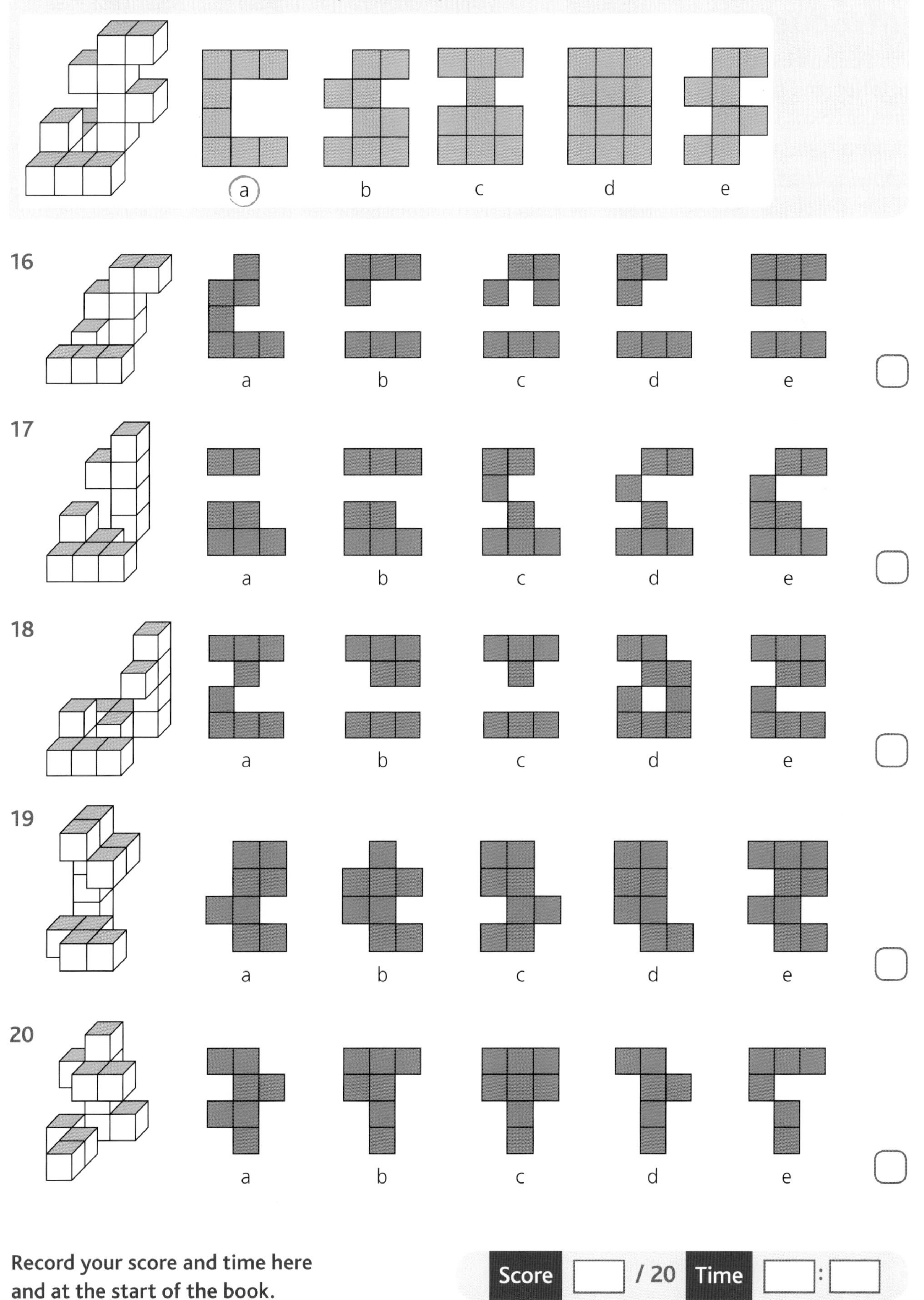

Record your score and time here and at the start of the book.

Score / 20 Time :

3 Links of position and direction

Introduction

Position and direction questions are linked to the work you have done in Maths on **reflection, rotation** and **translation**. Some types of question concentrate on testing just one of these areas of Maths at a time (very much like the 2D and 3D question types at the end of Chapter 1). However, you will also see position and direction used in *Most unlike, Matching features* and *Applying changes* questions where the position of some elements is important.

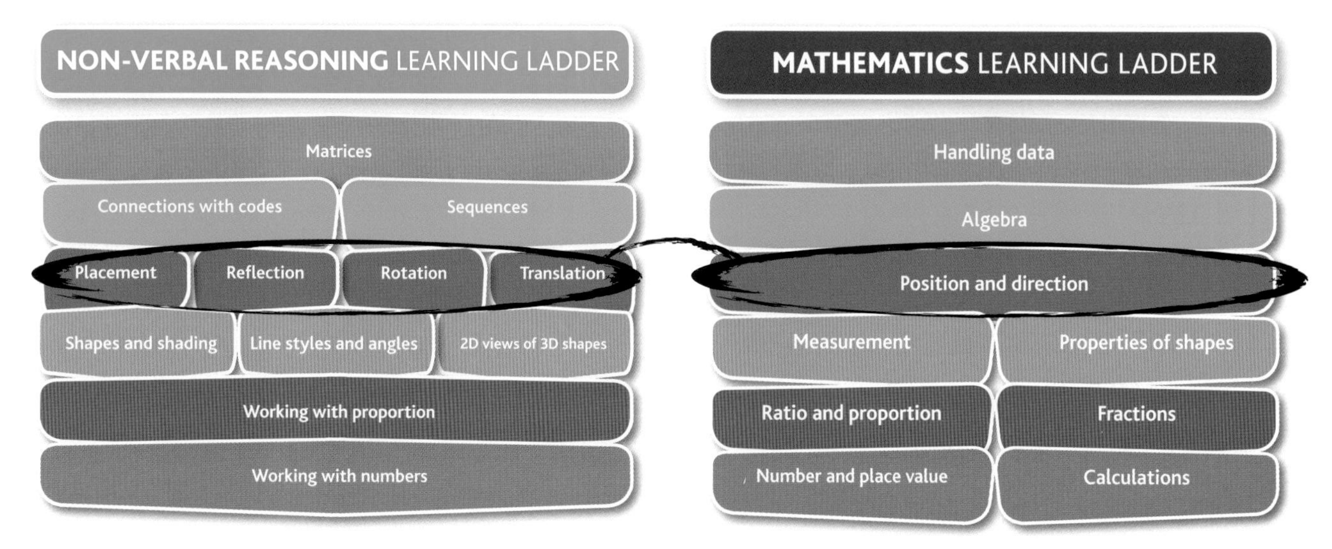

Chapter 3 moves up to step three on the ladders – so you can see you have learned a lot of skills!

Compare the steps on both ladders to see how **placement**, **reflection**, **rotation** and **translation** are connected to **position** and **direction** in Maths. All the questions in this chapter will relate to these skills, although you may find some distractors involving skills from steps one to three to make the questions more challenging.

Advice for parents

Some children find the skills explained this chapter very challenging, while others find them relatively easy. If your child finds these skills difficult, patience is the key together with a lot of support and a range of techniques (which are explained in the following pages) to help develop strategies to solve the problems.

As with 2D and 3D shapes, skills in reflection, rotation and translation are testing both the ability to hold an image in your head as well as the ability to perceive it from a different angle or position. Even if a child is very quick at solving the logic problems introduced in Chapter 2, the memory skills in Chapter 3 are slightly different and some people always struggle with them.

There are many electronic memory games on the market that can help improve response rates to this kind of question. Working with jigsaws is also excellent practice as they involve spotting rotated images and then fitting specific shapes – the skills in this chapter in a nutshell!

Chapters 4 to 6 explore many more examples of these spatial skills, with additional tips to help children increase their confidence.

Thinking skills and games

The following games can help to develop the skills featured in this chapter.

Rotation Pelmanism

Choose three photos to use for this Pelmanism game to play with two or more people. It should be clear which way is the right way up for all the photos or the game will not work – photos of people and animals work well.

Create a template for the cards, using the layout shown here. You will need to create a separate sheet for each of the three pictures. Each card should show a specific degree of rotation in 45° steps from 0° to 315°.

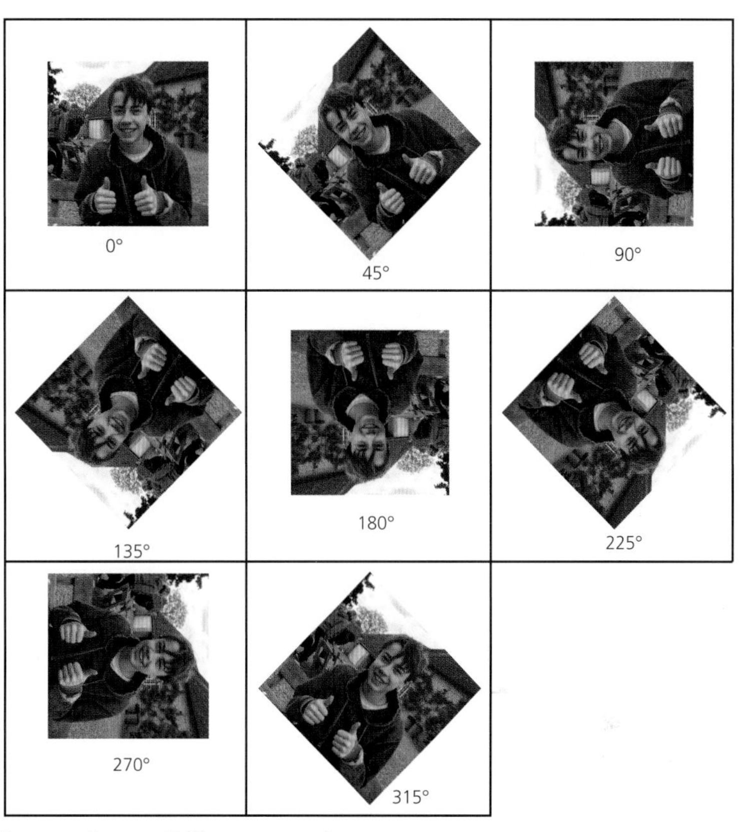

Print out each sheet on card **twice** and cut them up to make 48 cards (it is important that the cards are the same size, otherwise you may be able to remember cards from their shapes).

Place the cards face down and take it in turns to lift up pairs to find two that match, placing the cards back down in the same place if the cards are different.

A player finding a matching pair keeps the pair on their own pile.

The game ends when all the pairs are found. The winner is the person with the most pairs of cards.

You could also draw patterns using shapes in a word-processing program that look like shapes in Non-Verbal Reasoning questions for this game if you want more practice in spotting abstract shapes from different angles.

Split personalities

Find a face-swap app and practise your translation skills.

Look for a photo showing a group of people and try to guess which faces are of similar shapes. Then use the program tools to select the face from one person and see how well it matches onto the shape of the other person's face.

Happy snapping!

Placement

Skill definition: Find patterns where the position of the elements change, or show direction.

- You have seen that there can be many different elements in a Non-Verbal Reasoning question. Sometimes the **position** in which they are placed can also be important.
- The common features you may come across are:
  - patterns made up of groups of shapes
  - the direction in which a shape or arrow is pointing
  - the position of shapes in relation to each other (such as next to a side or vertex) and whether they are touching or separate
  - shapes that overlap and create layers. These generally stay the same shade although, in more difficult examples, the shading may change as well.
- Sometimes the changing feature is a space rather than any of the elements in an image. Here the gaps in the squares are all facing outwards.

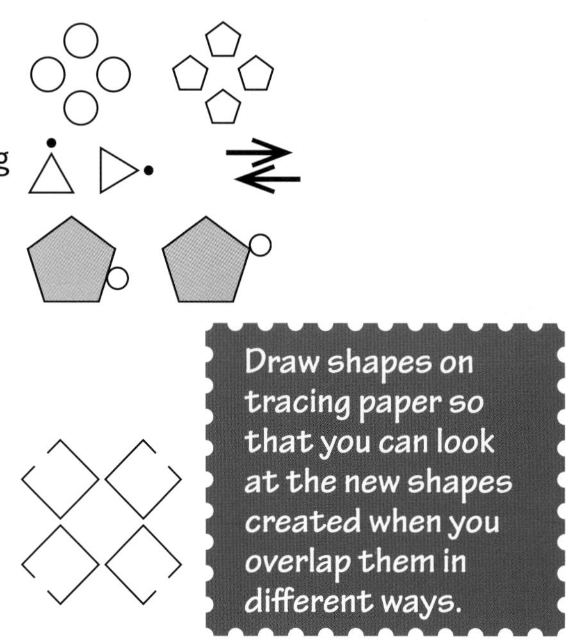

Method

Look at the two sets of pictures on the left connected by arrows and decide how the first pictures have been changed to create the second. Now apply the same rule to the third set and circle the letter beneath the correct answer from the five options.

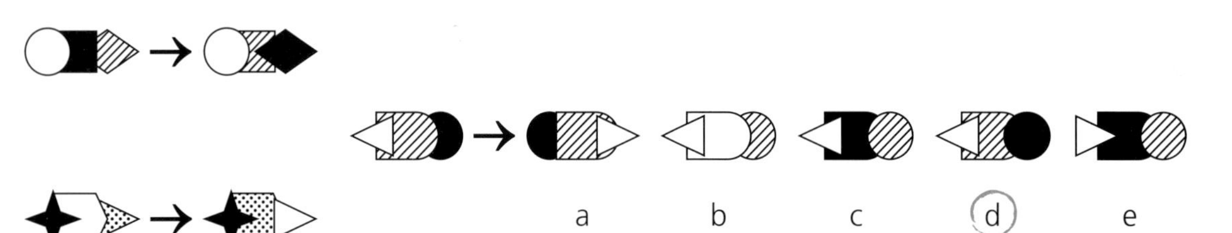

- This is a variation on the *Applying changes* question on pages 30–31, with an extra example.
- Start by comparing the first pair of pictures.
  - The white circle on the left does not change
  - The black and striped shapes change in shading and also in position – the middle shape overlaps the right-hand shape in the first picture but the right-hand shape overlaps the middle shape in the second picture.
- Now look at the second set of pictures.
  - The black star on the left does not change.
  - The middle and right-hand shapes appear to change shading and shape. However, the rule used to change the bottom set of pictures is the same as in the top set, even though it is more difficult to see what is going on with the polygon and the triangle.
- Finally apply the same rule to the picture on the right.
  - The white triangle will not change, which means you can reject options **a** and **e**.
  - Looking at the two shapes on the right of this picture, you know that the right-hand shape will overlap the shape in the middle. Both **c** and **d** could still be correct, but not **b**.
  - You also know that the shading *and* the position swap, so option **c** must be the correct answer.

Train

1 The set of squares on the right is made up of one large 'transparent' square and a number of other smaller squares that are shaded either black or white.

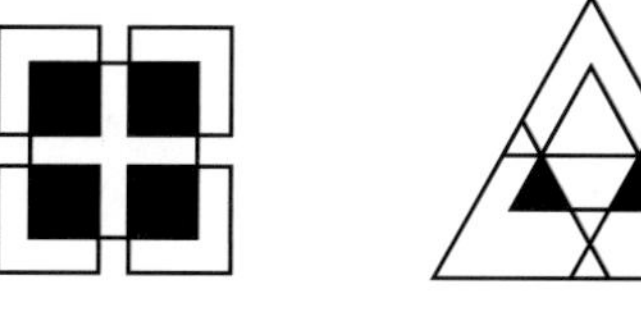

(a) How many squares are there in this picture? ______

(b) In the set of triangles above, the white triangles are all 'transparent' and the same size.

What is the total number of triangles, both black and white, in the picture? ______

Try

Look at the two sets of pictures on the left connected by arrows and decide how the first pictures have been changed to create the second. Now apply the same rule to the third set and circle the letter beneath the correct answer from the five options.

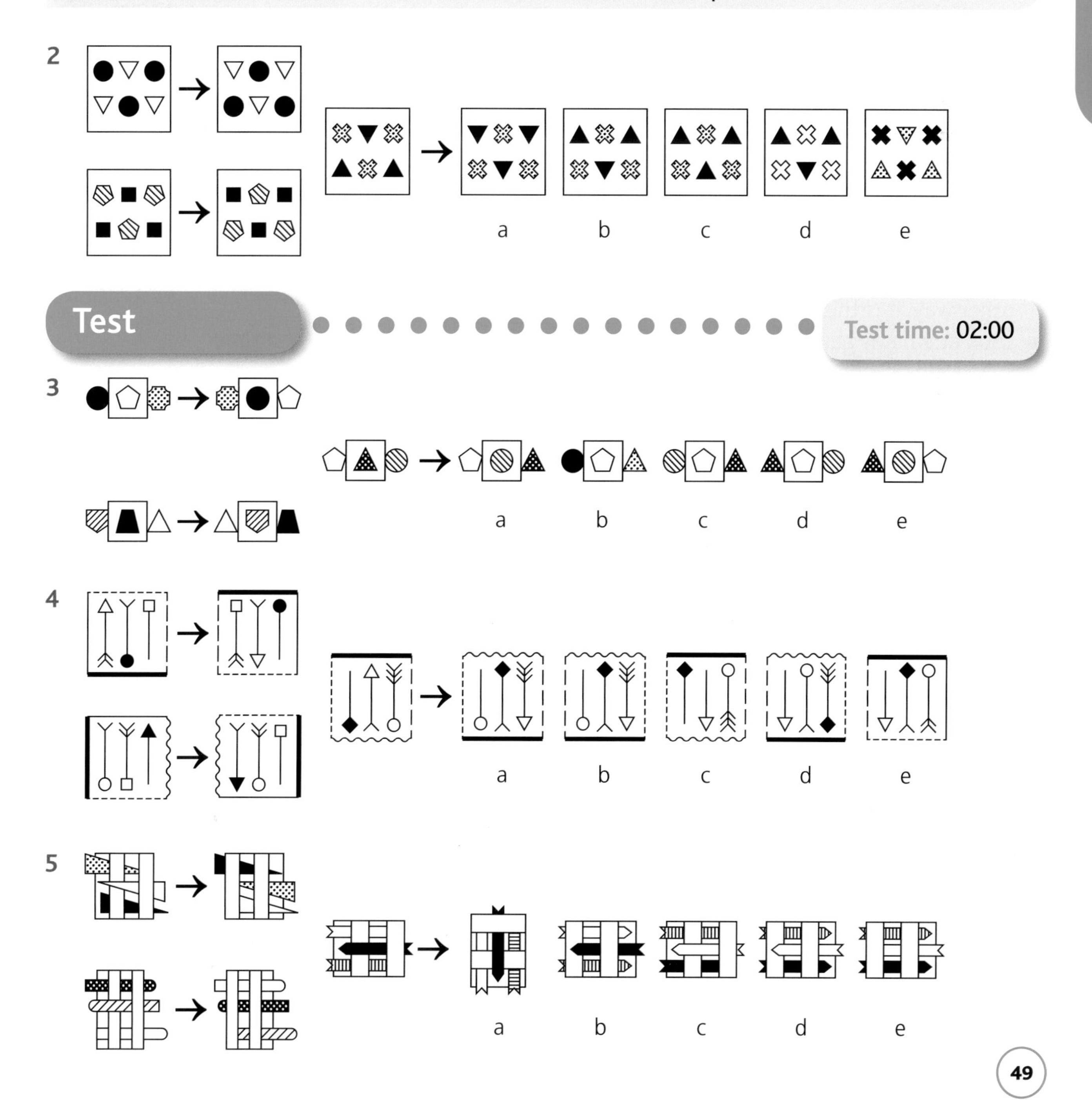

Symmetry and vertical reflections

Skill definition: Identify images that have lines of symmetry and vertical reflections.

- **Symmetry** is often seen in *Most unlike* questions, such as question 2 on page 33.
- Sometimes pictures have **more than one line of symmetry**. If the shapes look symmetrical, this is often a clue to the answer. Practise questions like these using a **mirror** (placing the mirror where the blue lines are shown on this example).
- Some question types test just one skill, such as vertical reflection, where you are asked to **match a single image** (as in the worked example below). Also look out for reflection questions where the **shading changes** – see the *Try* questions on page 51.

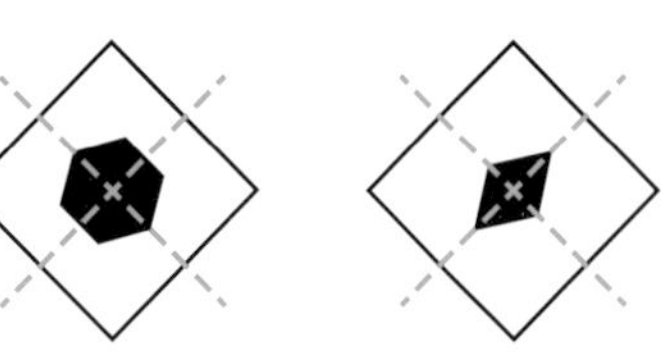

Method

The picture on the left is reflected in a vertical mirror line and is represented by one of the pictures on the right. Circle the letter beneath the correct answer.

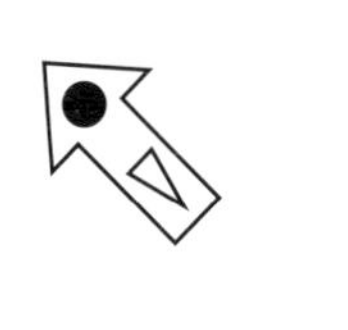

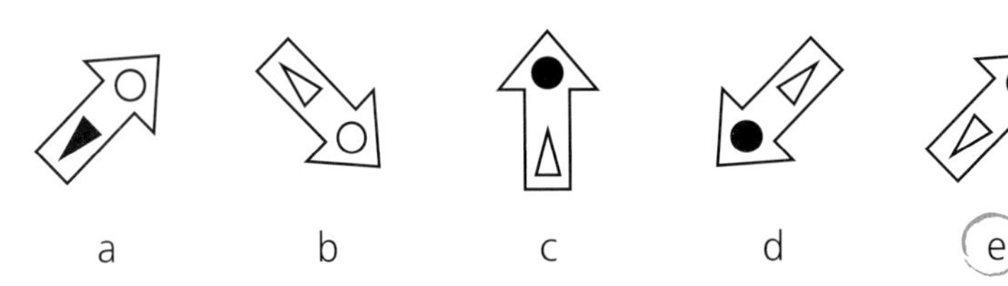

- As the question is about reflection, **nothing will change apart from the position** of the elements. The shading, shape and line styles will not change.
- Look at the arrow on the left. Imagine that the **ink is still wet** and you are folding the paper down the dotted line. When you unfold the paper, an identical image will appear pointing the other way. It is this 'reflected' image that you are looking for.
- Look at what is happening to the different features in the picture one at a time.
  - First look for any answers you can reject by **comparing the shading**. You are looking for a black circle and a white triangle. So options **a** and **b** must be incorrect.
  - Next, check which way up the pictures are. You know that the picture you are looking for must be **the same way up** (think about the wet ink image created when the paper is folded). So option **d** must also be incorrect as it is pointing downwards.
  - Now you are left with options **c** and **e** which look almost the same. Look at the **separate elements** to see if you can work out what is going on. Both circles look the same, so look at the triangle. We know this **must be pointing the same way** within the arrow as in the original picture, so option **e** is the answer because the triangle is pointing downwards.
- Check your answer using a mirror. Remember, in **vertical reflections** the shapes *always* stay the same way up; they just **flip from side to side**!

Scan or photograph some patterns, then flip them using the rotation tool in a word processing program. Look at shading, angles of shapes and lines. Print your images, draw in mirror lines and join up the vertices. Each vertex should be joined to the 'mirror line' by the shortest possible distance, as shown here.

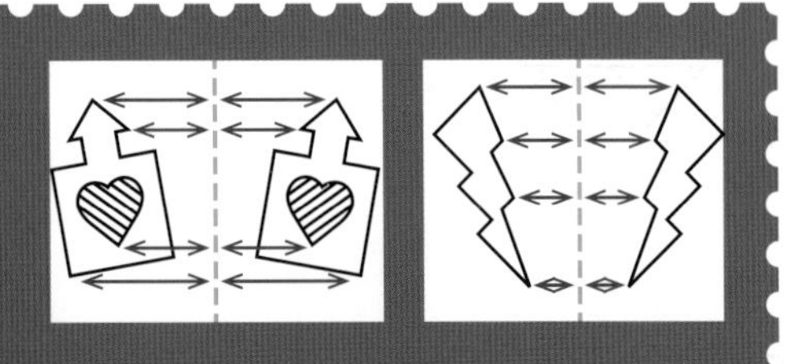

Train

1 Colour all the triangles black in rows 1, 3 and 5. Colour all the squares black in rows 2 and 4.

(a) What shape is formed by these coloured shapes?

____

(b) How many lines of symmetry are there in this coloured shape? ______________

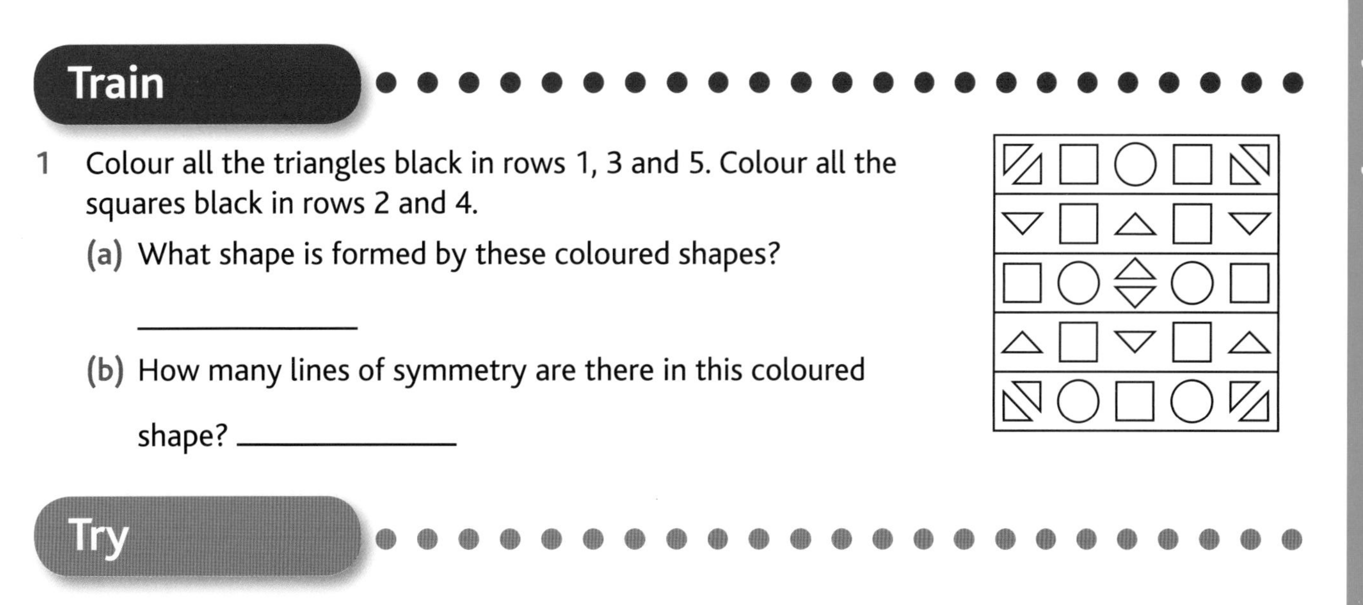

Try

The shapes on the vertical line are made out of paper that is black on one side and white on the other, then pasted onto a window. One of the answer options shows the same pattern viewed from the other side of the window. Notice that the colour reverses when viewed from the other side. Circle the letter beneath the correct answer option.

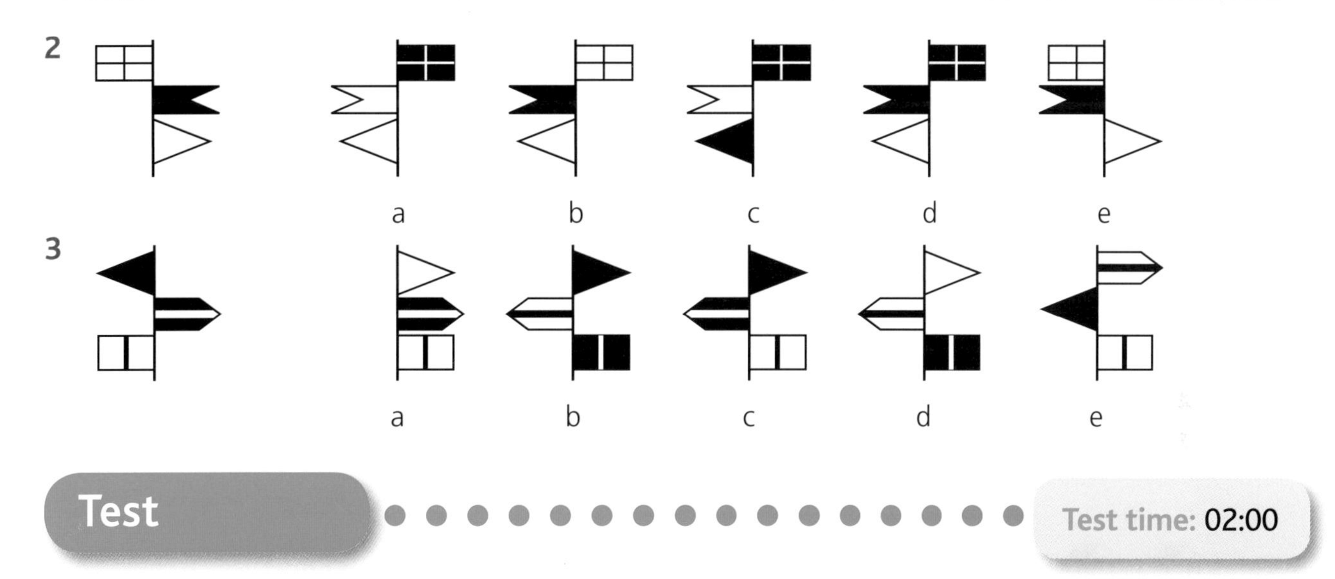

Test

Test time: 02:00

The picture on the left is reflected in a vertical mirror line and is represented by one of the pictures on the right. Circle the letter beneath the correct answer.

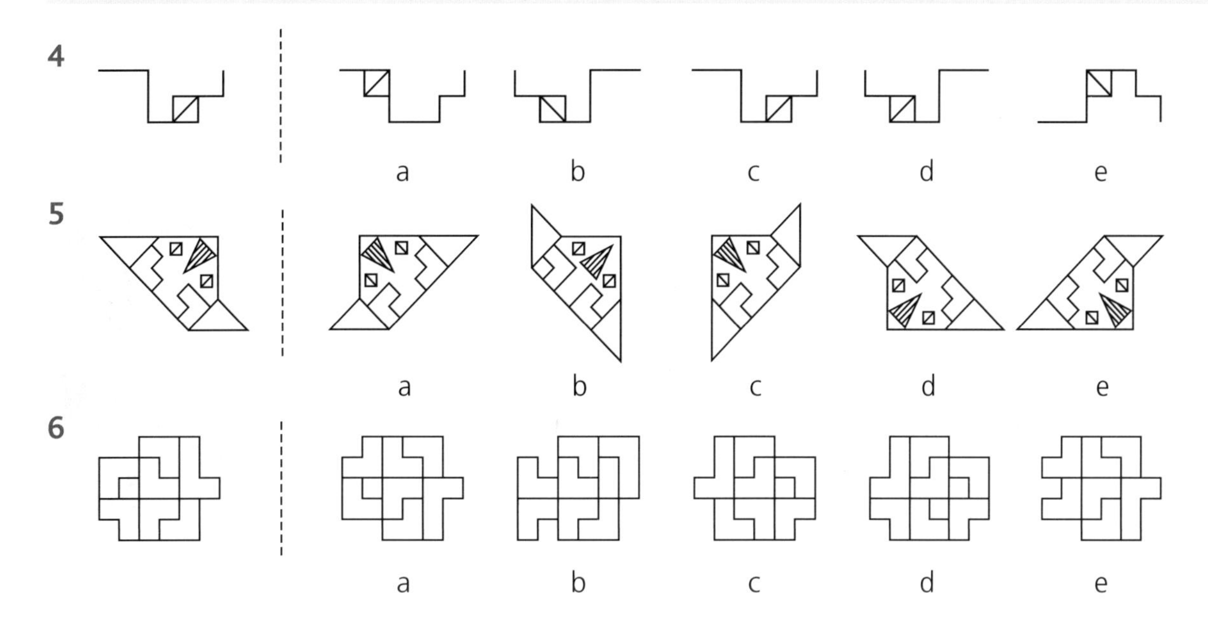

More reflections

Skill definition: Identify pictures that have been reflected in a horizontal or diagonal line and follow multiple reflections.

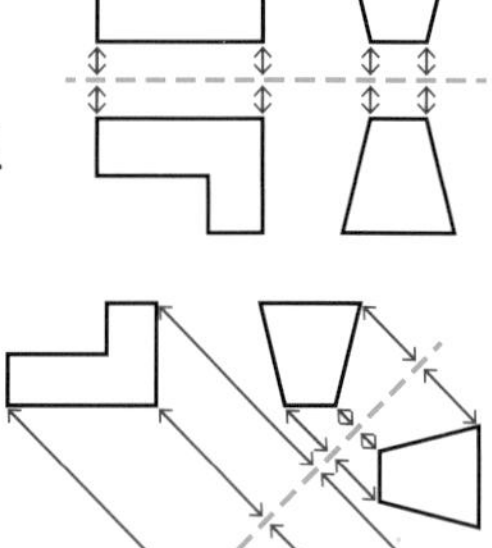

- In **horizontal** reflections, the shapes appear upside down. The vertices are all the same distance away from the mirror line on both sides.
- **Diagonal** reflections are harder to follow as the shapes seem to move around. If you think you have found a diagonal reflection, draw in the mirror line to check if the vertices are the same distance away from the line on each side as shown here.
- Thinking about the 'wet ink' rule that you used for vertical reflections can also help you work out what is going on. The questions on this learning spread all use **folds** and all test your **skills in reflection**.

Practise reflections by drawing the pictures on tracing paper and folding along the reflection lines. You can then see what the reflection will look like on the back of the tracing sheet. If you are given spare paper in the test, press hard with a pencil on the sheet and look at the impression on the reverse.

Method

The square given at the beginning is folded in the way indicated by the arrows, and then holes are punched where shown on the final diagram. Identify the answer option which shows what the square would look like when it is unfolded. Circle the letter beneath the correct answer.

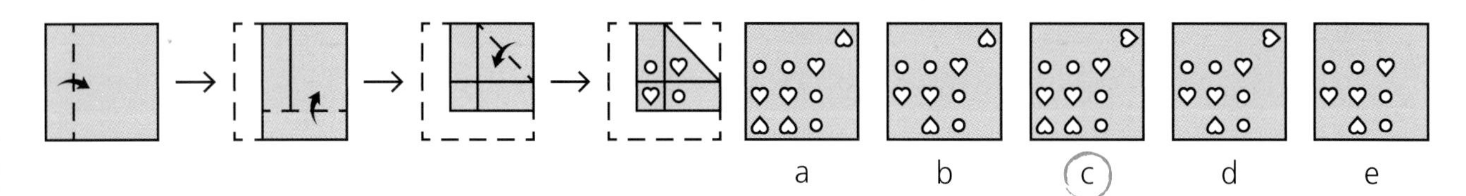

- Think about the **folds** as **lines of reflection**, imagining the shapes are an equal distance from the fold (or mirror line). Imagine unfolding the shape, **working backwards from the last fold** made.
- Colouring in the folded sections like this can help when practising these questions to make it clearer which layers overlap.
  - The **diagonal fold** is coloured orange. This was the last fold made so, the first one to look at. The fold does not overlap the first two folds, so the hole will have only gone through the paper **twice**.
  - The hole in the **bottom layer** of the paper will always be in the **same place** as shown in this final picture. The unfolded corner will be **reflected diagonally** in the fold, so this heart will point to the right. Options **a** and **b** show a horizontal reflection, so you can reject these. Option **e** misses out this reflection completely.
  - The second fold to look at is the **horizontal fold** upwards. This is the purple and blue section. The blue section overlaps the bottom layer and the purple section also overlaps the vertical fold so there are four layers of paper here.

- You know there will be a heart and circle in the **bottom layer** in the **same position**. You are also looking for a **horizontal reflection** of the heart and circle **below**. Both **c** and **d** show this reflection, so look at the final fold.
- The purple section shows that the heart is also punched through the two layers made by the vertical fold. Flip the shapes right to left to work out the reflection.
- Option **c** is correct because the hole went through **four layers** so you are looking for **four hearts**.

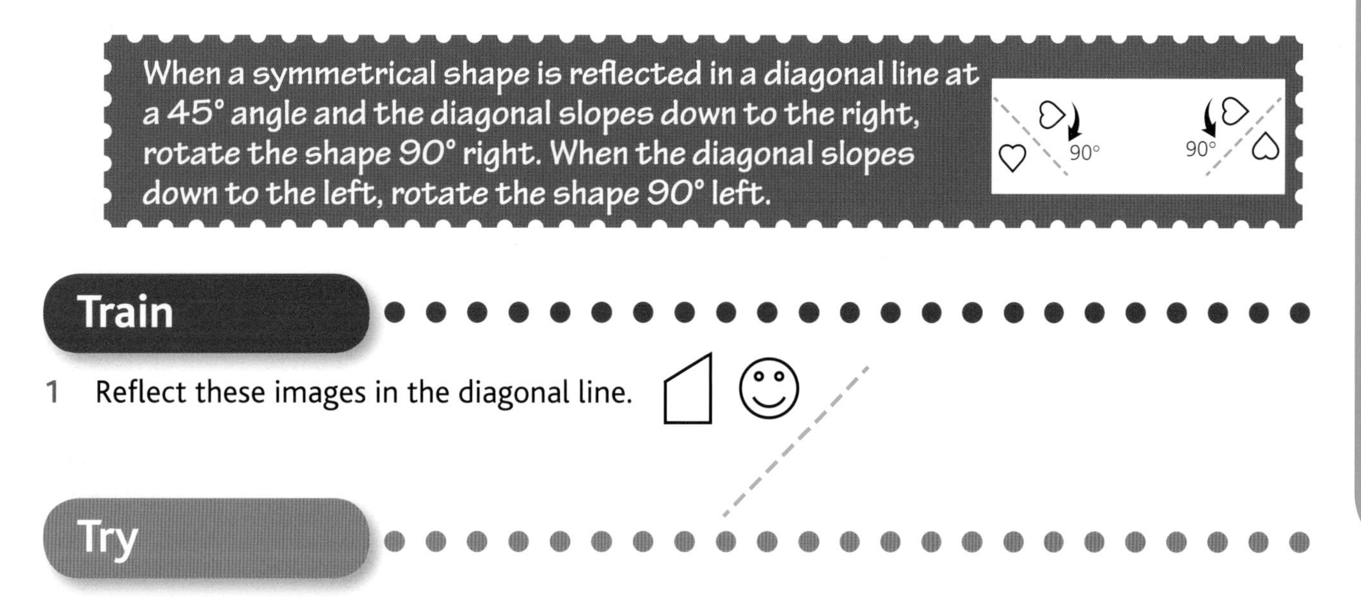

The square given at the beginning is folded in the way indicated by the arrows, and then holes are punched where shown on the final diagram. Identify the answer option that shows what the square would look like when it is unfolded. Circle the letter beneath the correct answer.

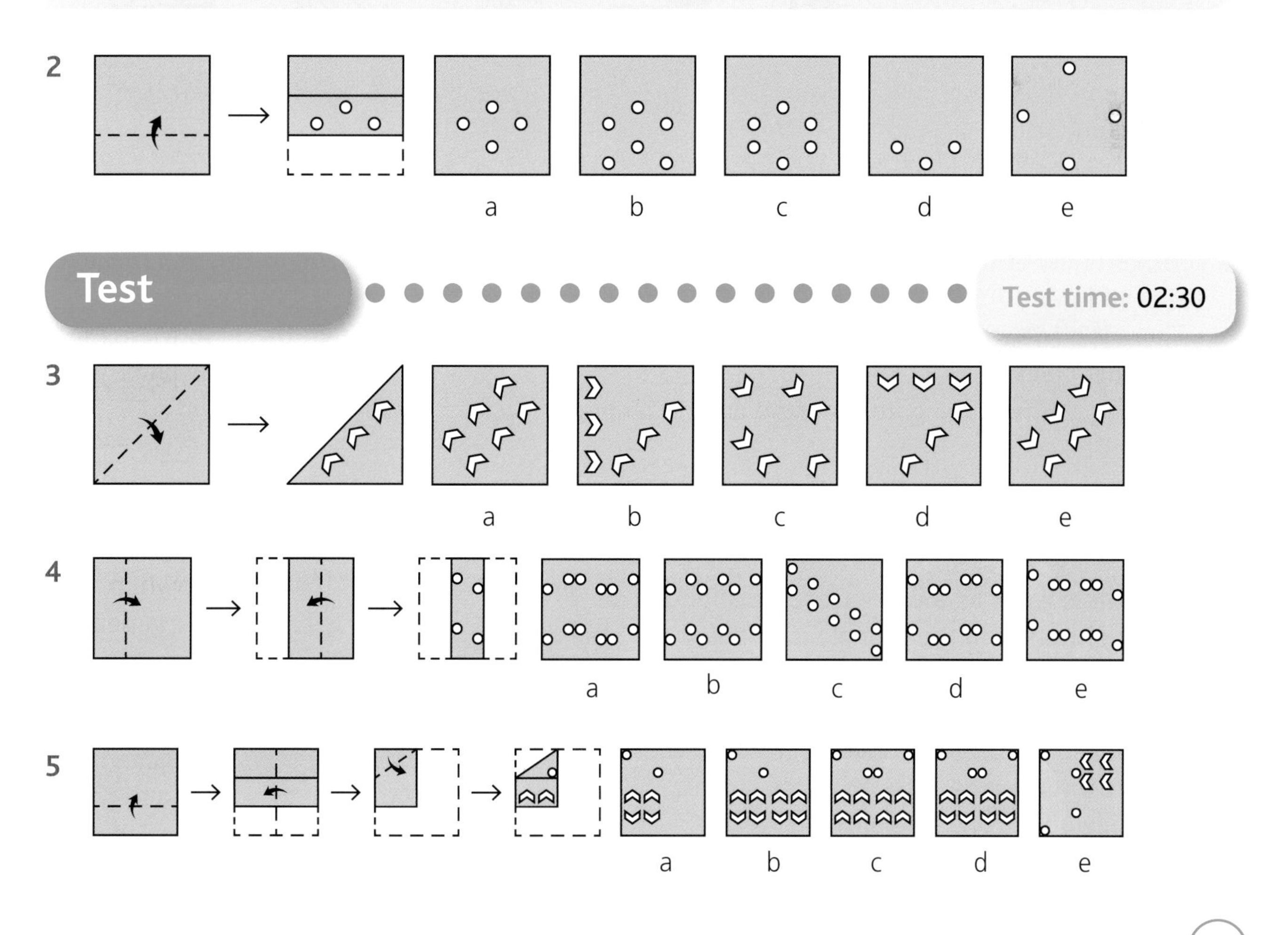

Rotations with 2D pictures

Skill definition: Identify 2D pictures that have been rotated when the degrees given are in 45° steps and when the degrees of rotation have not been specified.

- If you find rotations difficult to spot, there is nothing wrong with **rotating the page** to help you to work out what is happening. You have probably seen your parents do this with a map!
- You may be asked to rotate a simple image in **45° steps**, and the **Pelmanism** game on page 47 will help you to spot these rotations. Practise with the *Try* questions on page 55.
- As most rotation questions **do not give degrees**, look for:
  - angles on lines
  - shapes giving clues to a change in direction
  - shading giving clues to direction
  - small shapes that give a clue to the position oflarger shapes.

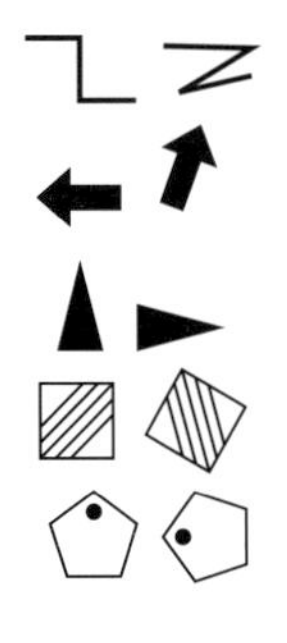

- Rotations also appear in *Applying changes* questions and in the complex question types in Chapters 4 to 6.

Method

The picture on the left is rotated and is represented by one of the pictures on the right. Circle the letter beneath the correct answer.

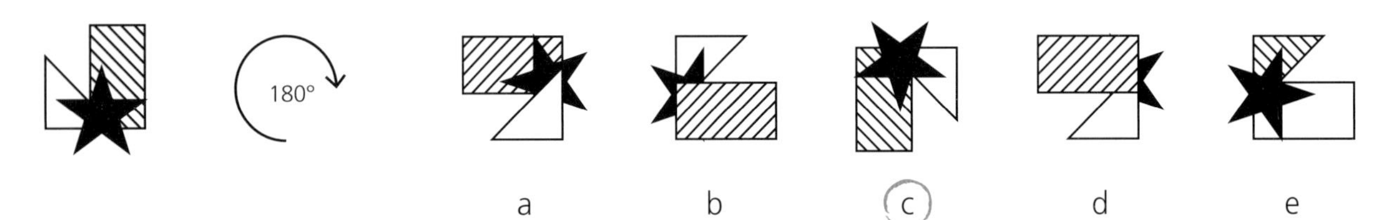

- In these rotation questions, the **answer option will be identical** in size, shape and shading to the picture on the left. The only change is how much the **entire image rotates**.
- Look carefully at the features of the picture before attempting the rotation as this will make it easier to find.
  - There is a triangle, a rectangle and a star. All the answer options contain these, so there must be other clues.
  - Looking at the shading, the triangle and rectangle in the final option swap shading so **e** cannot be correct.
  - As the shapes overlap, this could be another clue. The star is at the front in the first picture. The only other pictures where it is in this position are options **c** and **e**, so **c** must be the correct answer as you have already rejected **e**.
- Check the answer by looking closely at option **c**. The right-angle in the triangle is on the outside, the rectangle is positioned with the longer sides parallel to the outside edge of the triangle, the diagonal shading works in the opposite direction because all of the shapes are upside down. The side of the rectangle and triangle that run in a continuous line runs through bottom of the star.

Diagonal shading rotated through 180° looks the same; reflected diagonal shading does not!

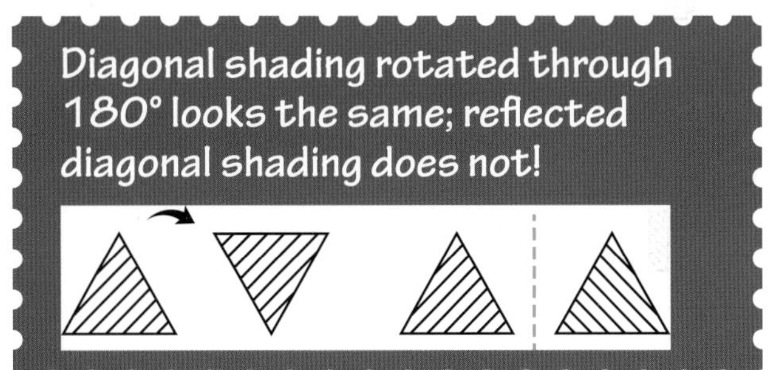

Train

1 Some of these shapes look the same when they are rotated the number of degrees shown in the table. Put a tick for the rotations that look the same.

Rotation	45°	90°	180°	225°

Try

The picture on the left is rotated by the number of degrees shown to give one of the pictures on the right. Circle the letter beneath the correct answer.

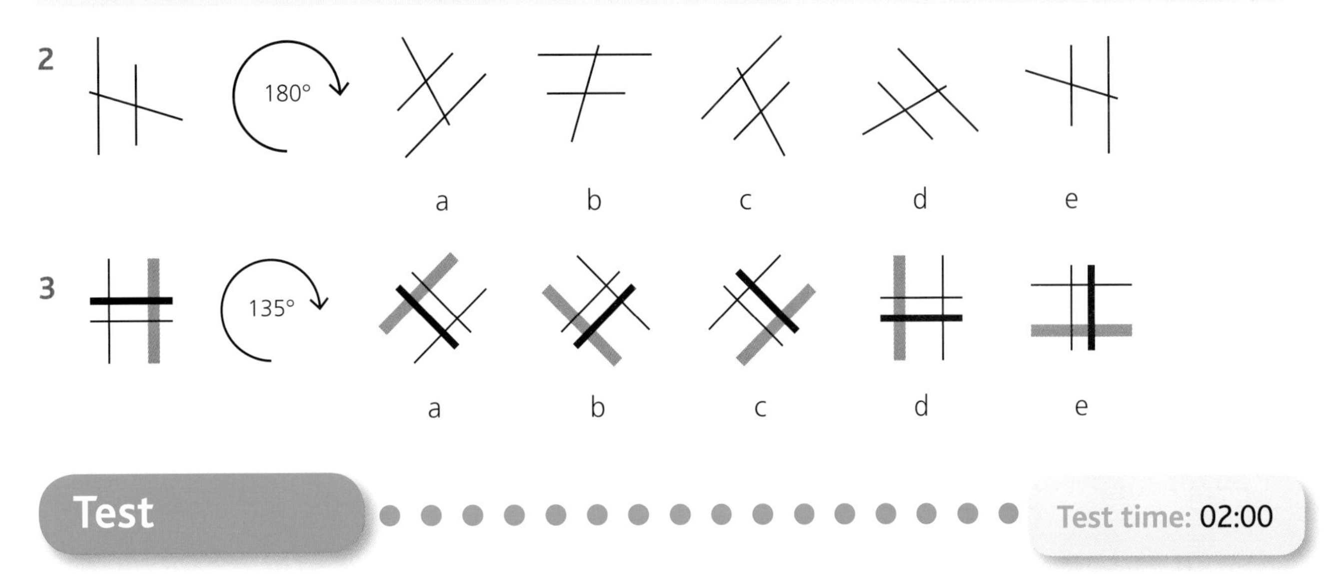

Test

Test time: 02:00

The picture on the left is rotated and is represented by one of the pictures on the right. Circle the letter beneath the correct answer.

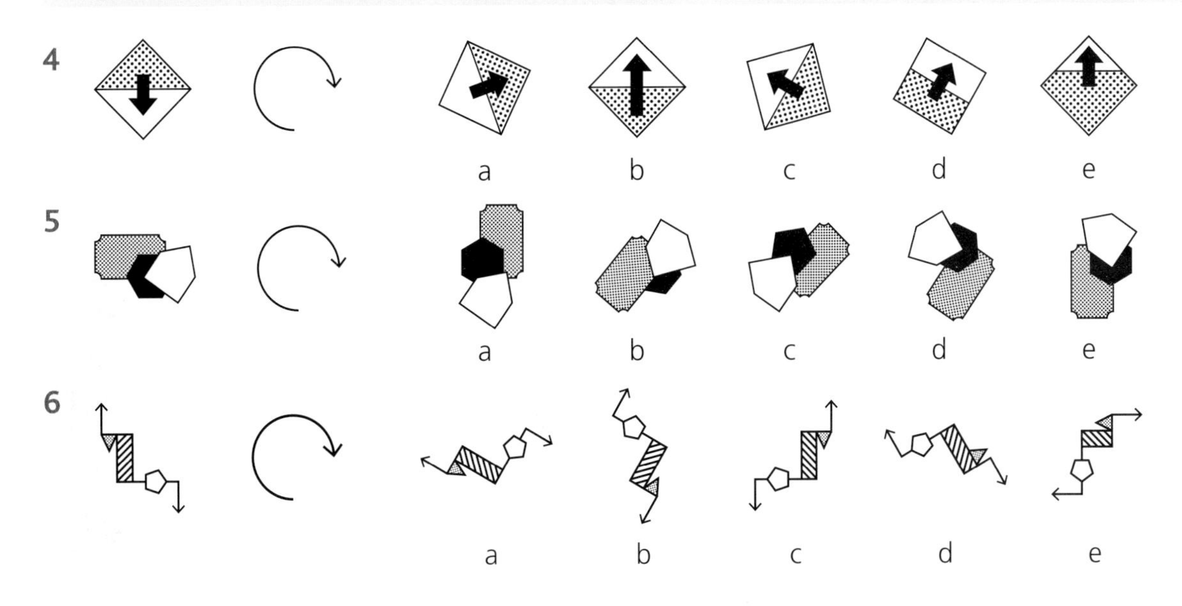

Rotations with 3D pictures

Skill definition: Identify 3D pictures that have been rotated, with and without patterns.

- *Rotations with 3D pictures* questions show either **blank cuboids** piled together, as in the example below, or **patterns on cubes** which work like the *Applying changes* questions you have already seen (and shown in the *Try* questions on page 57).
- 3D rotations can be tricky to work out, although the **blank cuboid questions** often give you a **selection to match up**. This means that, once you have matched the easier ones, there are fewer to choose from to complete the questions.

Method

Circle the letter next to each question that matches up with the rotated pictures **a** to **e**.

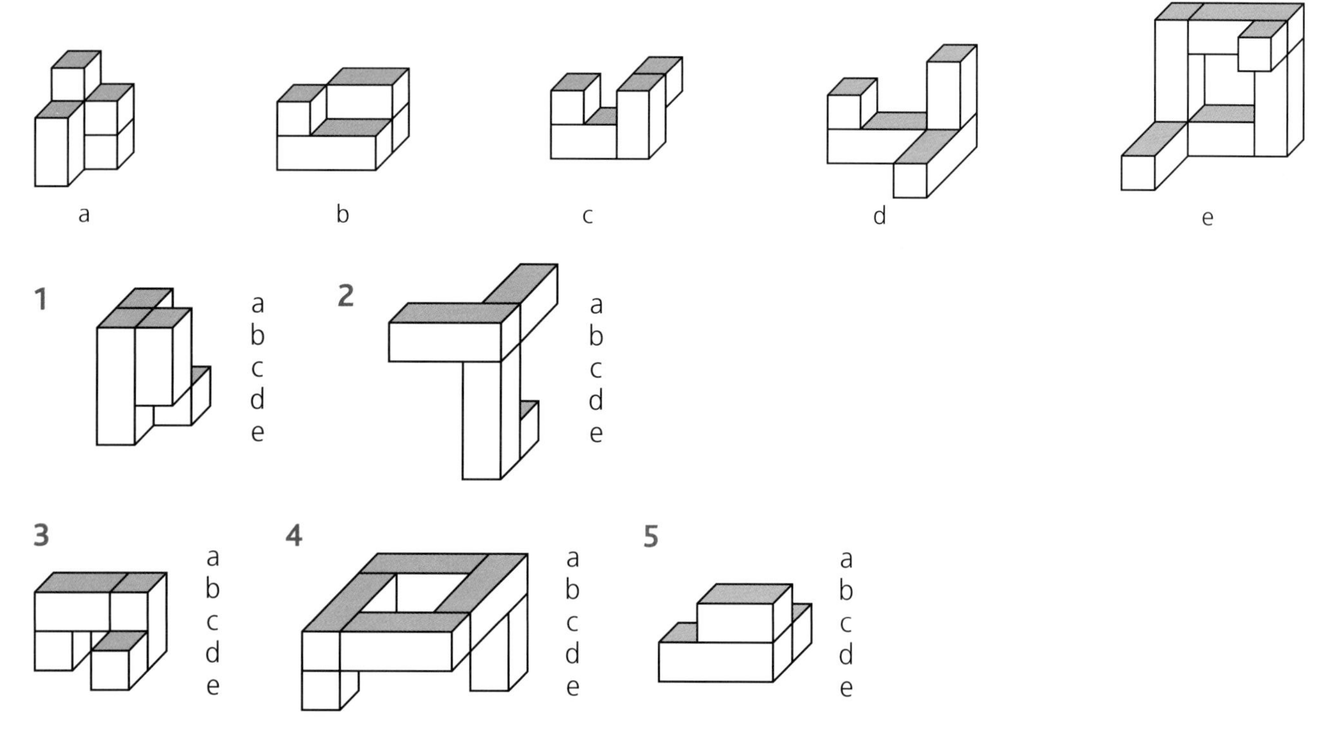

- These questions are really **counting** and **matching** exercises and can be solved in a similar way to the questions about plans on pages 40–41 by looking at how many cubes are **in a line**.
- Begin with an image that has clear features to find some answers quickly.
  - Question 4 has blank space in the middle and cuboids making up a square. Option **e** is the only question with this feature, so must be the correct answer.
  - Question 2 has a large L-shape on top. **Count** how many cubes **wide** and **deep** the L-shape is. Option **d** is the only answer with this feature.
- **Leave questions with similar features to the end**: Questions 1 and 5 look similar as they contain blocks of three cubes with two cubes at the front, so go on to question 3.
  - Option **c** seems closest with a backwards L-shape at the front made up of two double cubes. Looking at the right-hand side of the picture in question 3, you can see the same pattern at the back of this picture.
  - Now look at the rest of the question. There are also two single cubes forming two shorter Ls. Imagine picking the shape up and rolling it forward slowly. Can you see it matches option **c**? The picture has been rotated 180°.
  - Finally decide which of the final two questions has **most cubes visible** as this will give you **the most clues** (question 1).

- You will need to imagine **turning it around** to match a pattern you can see in the answer options.
- Turning the picture upside down and to the left puts the two and three cube blocks into the same position as option **a**, but you also end up with a small cube at the top which does not match. Therefore the answer must be option **b**.

Train

1 The cube dog can rotate in any direction, but always returns to the same position.

(a) How many degrees must he rotate to turn to face you? ________

(b) How many degrees must he rotate to look in the opposite direction to the one he is looking in now? ________

Try

The first cube has been rotated to a new position, shown after the arrow. One of the answer options shows the third cube rotated in the same way as the first cube. Circle the letter beneath the correct answer.

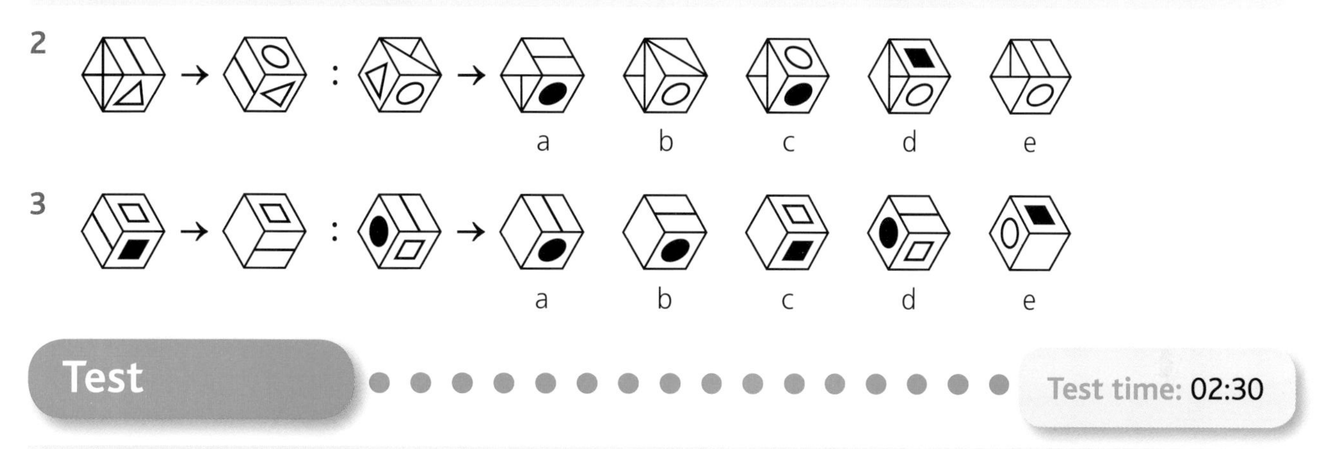

Test

Test time: 02:30

Circle the letter next to each question that matches up with the rotated pictures **a** to **c**.

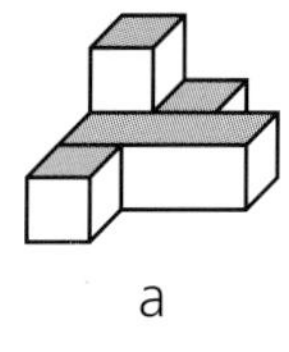
a

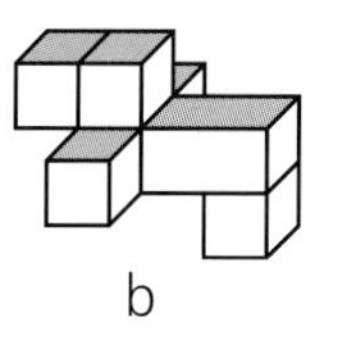
b

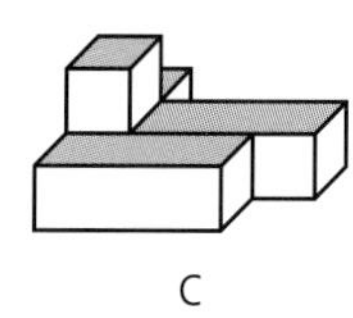
c

4 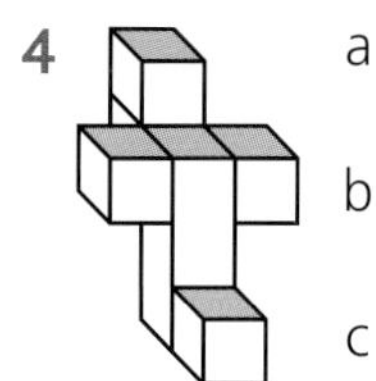 a b c

5 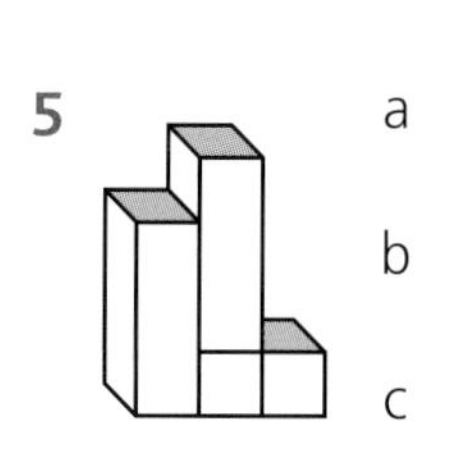 a b c

6 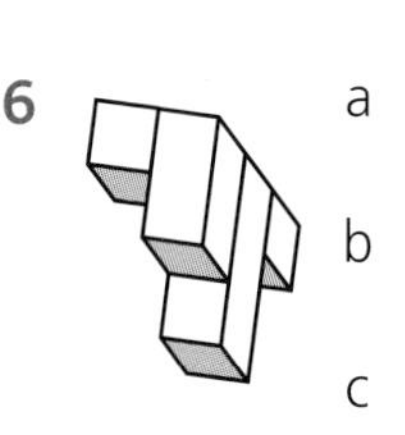 a b c

Translations with 2D pictures

Skill definition: Identify complex 2D pictures that have been translated or single 2D shapes that have been translated onto a more complex picture.

- **Translated** pictures only move from left to right or up and down; they **do not rotate**. Answer options often include rotations to try to trip you up, so watch out!
- Often the differences are subtle so you need to do some detective work to look for clues.
- There are two main types of *Translation* questions and both use the skills below:
  - **finding a simple shape** in a more complicated picture (see example below and *Train* questions)
  - **matching a group of shapes** to an outline or finding a set of dots in a pattern (see *Try* questions).
- The trick with these questions is to look for:
  - angles that match
  - proportions of shapes – check these do not change
  - the position of sides and lines within shapes – do they match or are they rotated?
  - features of the shapes themselves: the numbers of sides and curves.

Method

The small shape on the left can be found in one of the pictures on the right. It might be made up of one or more pieces. Circle the letter beneath the correct answer.

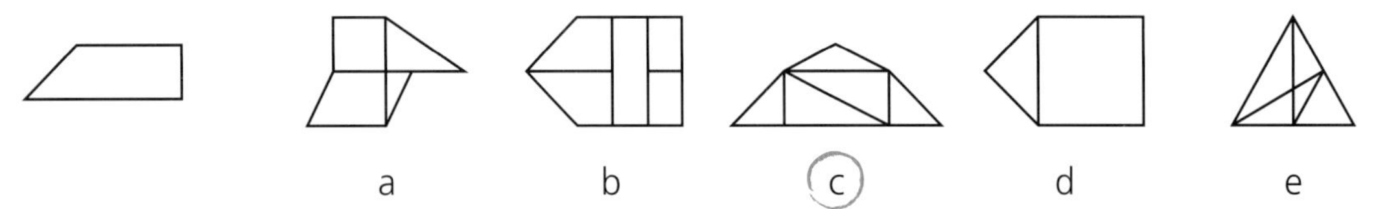

- Tell yourself before you begin that you are **not going to rotate** this shape. The shape on the left will always have some features that **stand out**, so decide on one and start looking through the answer options.
  - There is an **angle** 'pointing' to the **left** on a **horizontal** side that looks about **45°**. Option **a** has one angle pointing to the left on horizontal lines (in the two quadrilaterals on the left and in the triangle on the right) but all of these angles are larger than 45°. Option **d** has one angle but it is a right-angle, so you can reject **a** and **d**.
  - Now move on to another feature to narrow down your options. The shape has two parallel sides and is quite wide. Option **b** has a band running vertically through the middle of the picture, so the shapes on the left are too narrow. The triangle in option **e** is also obviously too narrow. So the answer must be **c**.
- **Check the answer** option you have chosen by looking at **all the features** of the shape on the left to be really sure your are correct.
  - There is a 45° angle on the left-hand side and two parallel sides.
  - The other feature of the shape is two right-angles. There are identical angles in option **c** so this must be the correct answer.

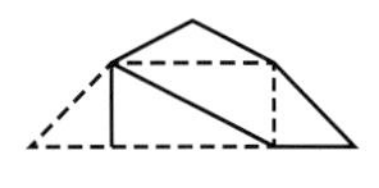

Train

1

(a) When this group of shapes is seen in silhouette (like a shadow on a wall) it makes a recognisable image. What is the image? ______________________

(b) Look up how to make hand shadow puppets on the internet and practise making silhouettes of your own!

Try

Look at the picture on the left. One of the holes on the right matches this picture exactly. Circle the letter beneath the outline that is an exact match.

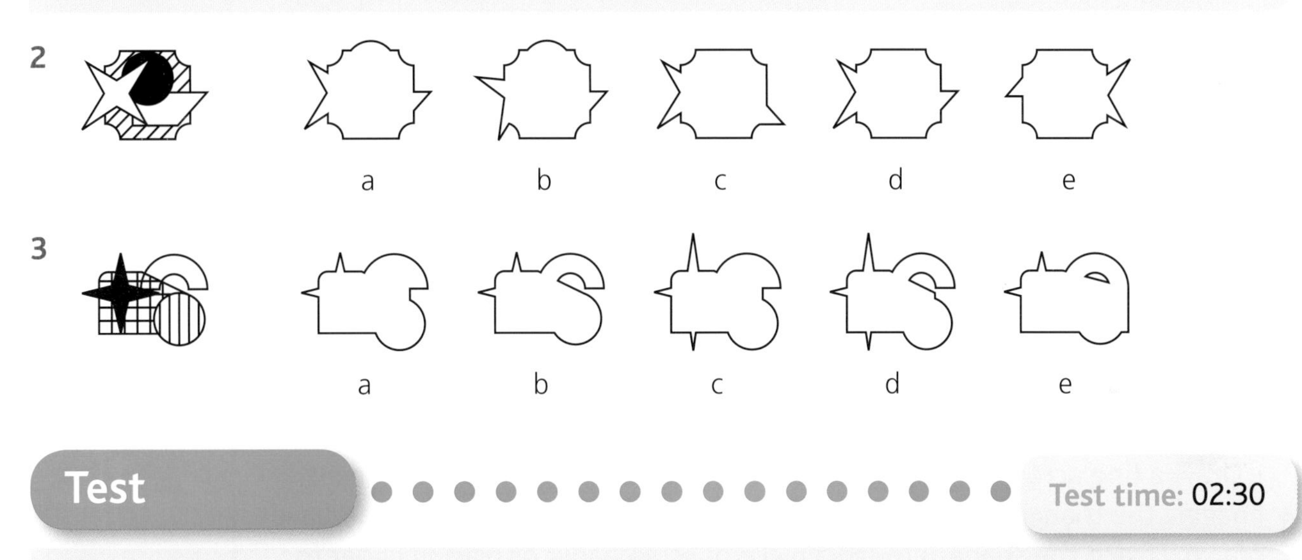

Test

Test time: 02:30

The small shape on the left can be found in one of the pictures on the right. It might be made up of one or more pieces. Circle the letter beneath the correct answer.

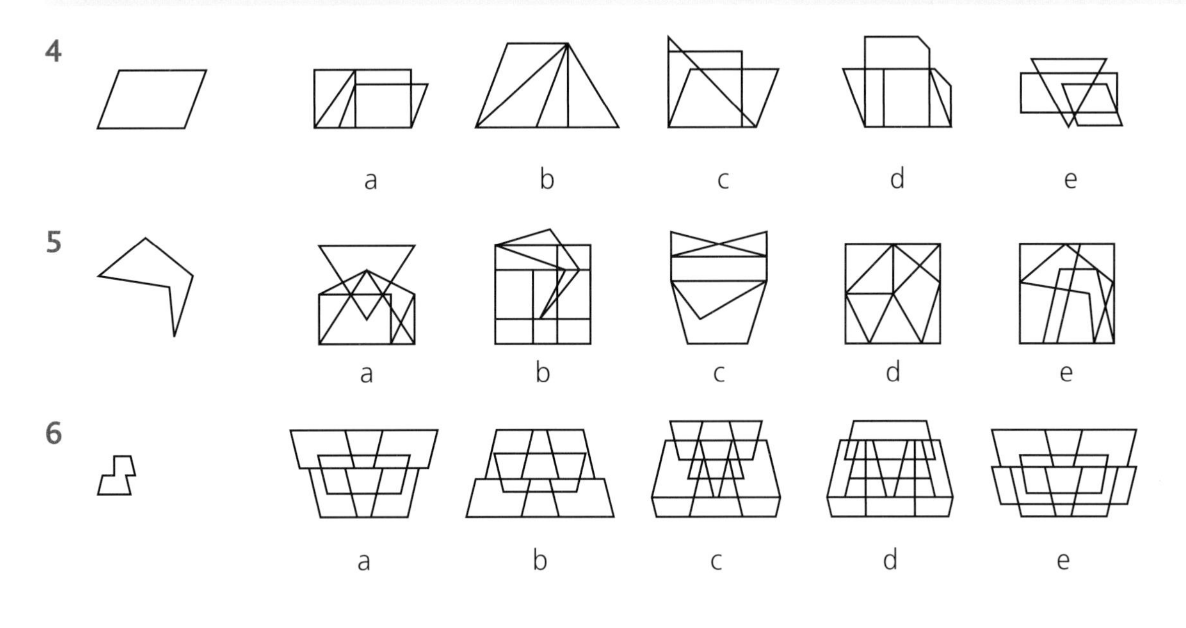

Translations and rotations

Skill definition: Identify 2D and 3D pictures that have been rotated and translated.

- Now you are familiar with *Rotation* and *Translation* questions, it is time to combine both of these skills.
- These questions ask you work out which group of shapes can be put together to make a different shape. Unlike *Translation* questions, **each shape** could be **rotated in a different way**.

Method 1

The shape on the left can be made up using three of the five smaller shapes on the right. Identify the **three** shapes needed and circle the letters beneath them.

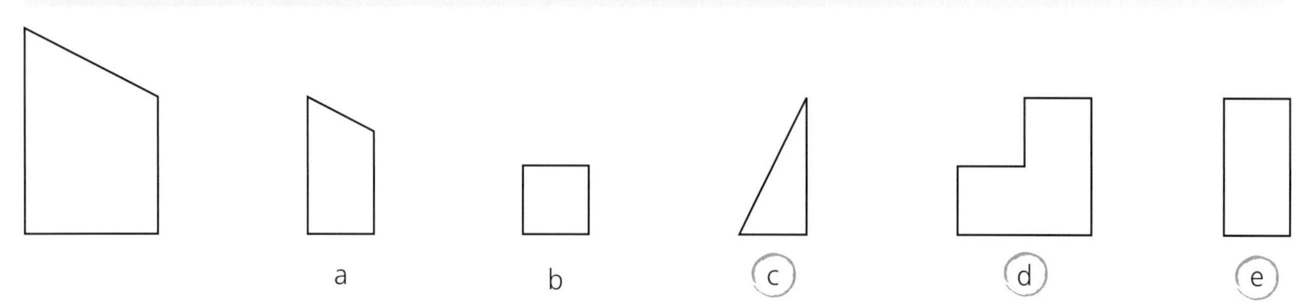

- These questions are unusual because you have to **circle three answer options**. It is important to **read the instructions carefully** to check what you are allowed or expected to do.
  - You are *not* being asked to scale shapes, so the **proportions do not change**.
  - You *are* being asked to combine shapes, so you must join **different shapes** together.
  - You have *not* been told the shapes are in the correct position, so you **can rotate** them.
- There are a lot of possibilities once you start looking at the answer options, so start with the big shape on the left to think about how you could **break it up**. Use your **knowledge of shape** from Maths and think the problem through.
  - This big quadrilateral can be divided into **a triangle** and a **square**.
  - **Draw a line** across the top of the square like this to examine the shapes.
- Imagine the triangle in option **c** rotated to fit on top of the triangle you have drawn. (**Trace the shape** through a piece of scrap paper or tracing paper to check the shape if you find this easier.) The triangles are exactly the same so **c** must be one of the correct answers.
- You are now missing two shapes that make up the **square** part of the quadrilateral.
  - Options **b** and **d** look like they might fit.
  - **Mark the length** of both shapes on the edge of a sheet of paper and match them to the source shape. They both fit the space so must be the last two answers.

If breaking down the complete shape does not give a clear answer, try marking the lengths of sides and size of angles onto paper. Then compare these measurements against the answer options.

Method 2

One group of separate blocks has been joined together to produce the pattern of blocks shown on the left. Some of the blocks may have been rotated. Circle the letter beneath the blocks that make up the pattern.

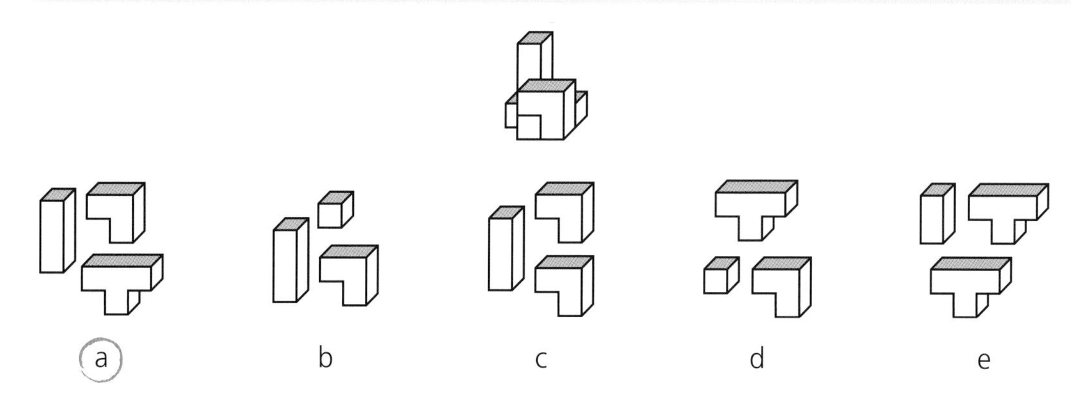

- There are usually some answer options that are obviously wrong in these 3D questions, so they are not quite as tricky as they may first appear.
- If the answer is not instantly clear to you, the fastest way to solve these problems is to work through the blocks **rejecting answer options** as you go along.
- Begin by looking at the **assembled pattern of blocks**. It is made up from cuboids joined together like the *Rotation* questions on pages 56–57.
- As you cannot see all the blocks clearly, **pick a block you can see** and begin there. You are going to **count the cubes** in the block to see if it appears in the answer options.
  - The only clear shape is the upside down L-shape at the front. This is made up from three cubes. Try drawing lines on the blocks if you find it difficult to see how many cubes you are looking at. Option **e** does not have an L-shape so we can reject this answer.

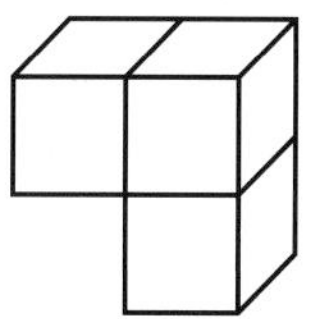

  - Circle one L-shaped block in the remaining answer options. This is to make sure you do not count it twice when looking at other blocks.
- As you cannot see any other blocks clearly, **look at the answers** and **pick a block** from there.
  - Start with the single block in options **b** and **d**. If using option **b** to make the pattern, this single block would go beneath the L-shape. However, the tall three-cube block could not then make the shapes at the back of the pattern. So this cannot be the correct option.
  - Move on to option **d**. The T-shaped block cannot make the shape at the back, but it could work in another way: **flip the T-shape backwards** and you have a shape that would fit the pattern at the bottom. This plan view shows the T in blue outline and the top of the L in green. The single block is not tall enough to complete the column at the back, so option **d** is also wrong.

- Look closely for the **most likely features** in the remaining options.
  - It looks like you need an L-shaped block, a T-shaped block and a tall block. Option **a** contains all three of these blocks.
  - See if you can fit the blocks together to **check the answer** – the three-cube block sits behind the T to complete the pattern.

When practising 3D rotations and translations, use sugar cubes or number blocks to help you work out what is going on.

Train

1

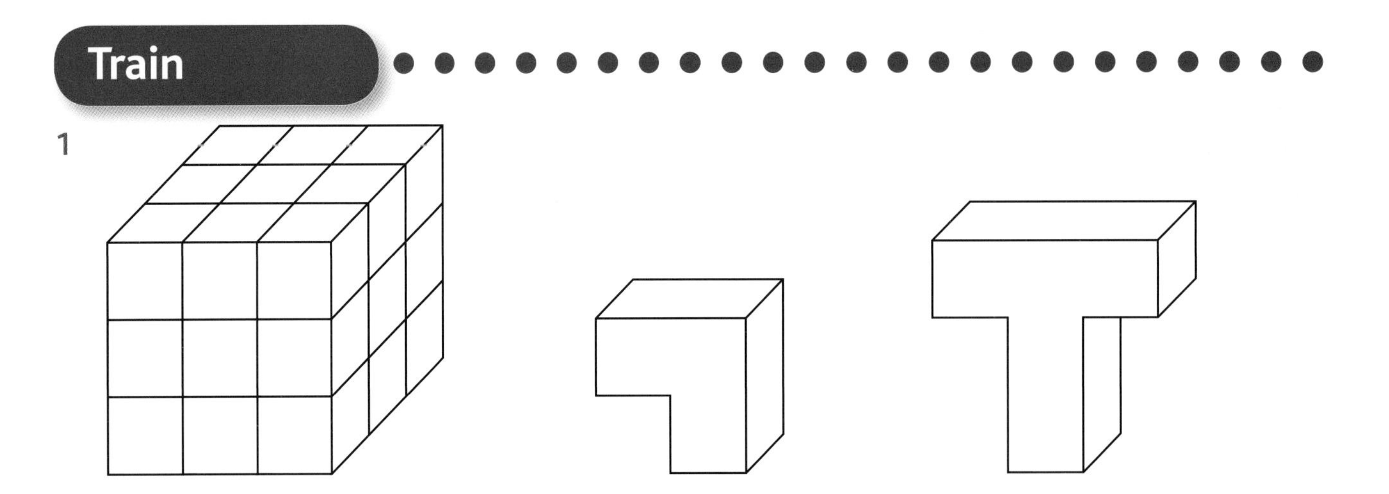

A large cube is made up of 27 smaller cubes. You have three of the T-shaped blocks and a number of L-shaped blocks.

If all three T-shaped blocks are upright, how many L-shaped blocks are needed to make up the cube? ______

Try

The shape on the left can be made up using three of the five smaller shapes on the right. Identify the **three** shapes needed and circle the letters beneath them. Example:

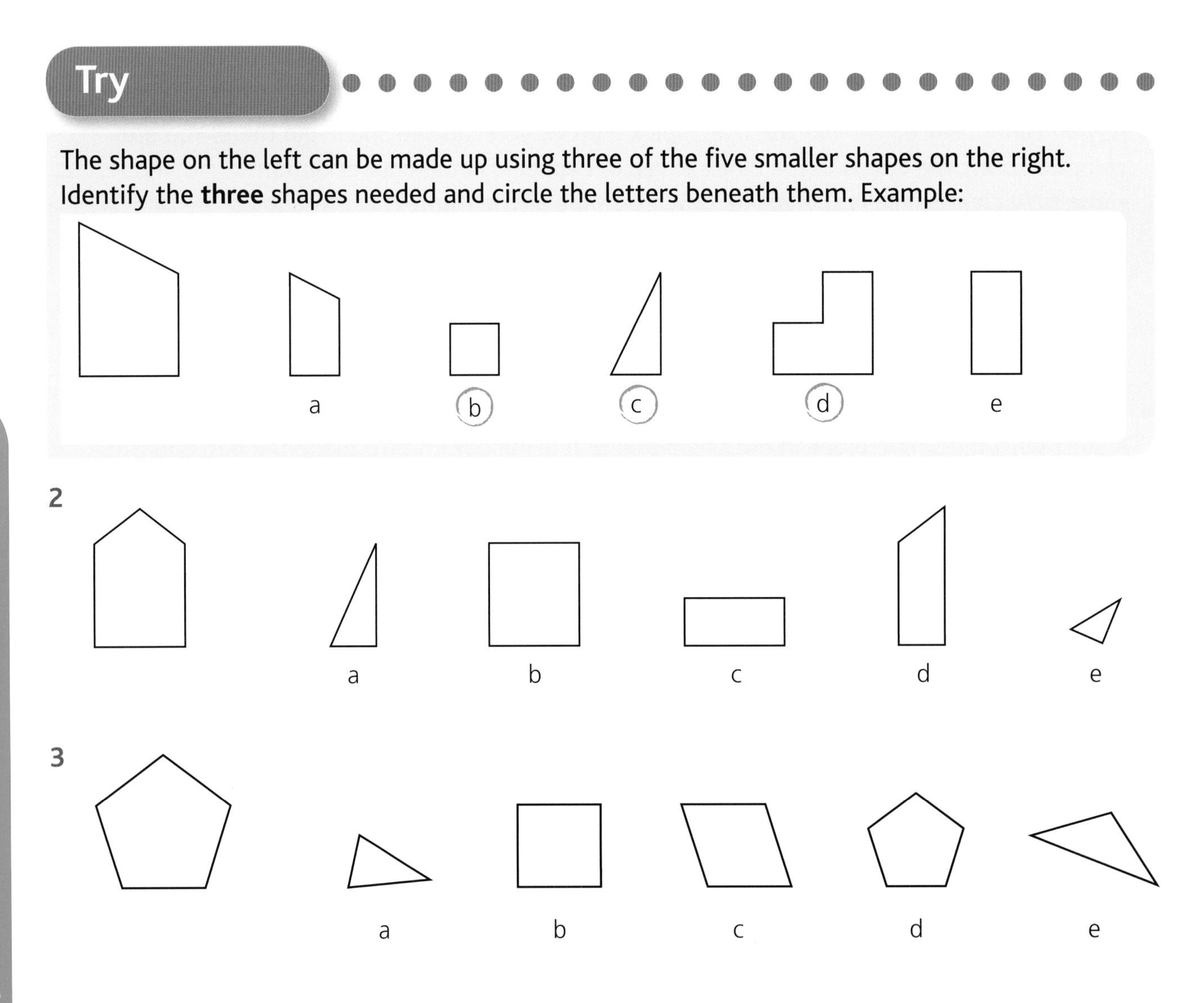

Test

Test time: 02:30

One group of separate blocks has been joined together to produce the pattern of blocks shown above them. Some of the blocks may have been rotated. Circle the letter beneath the blocks that make up the pattern. Example:

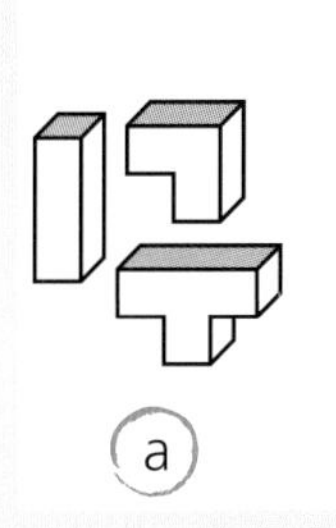
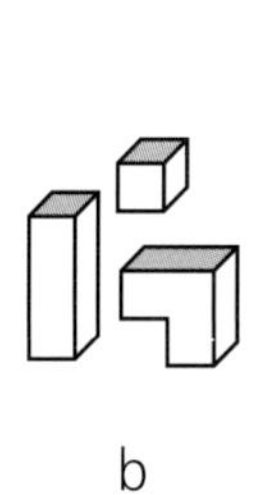
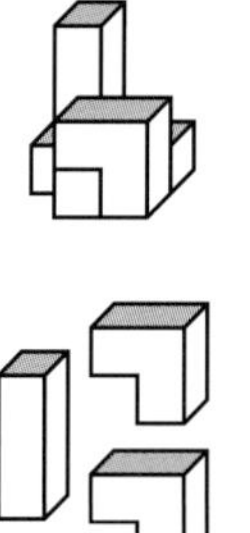
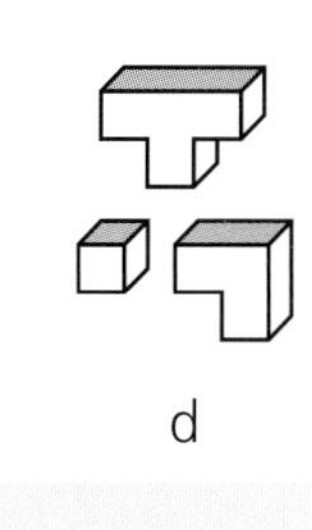
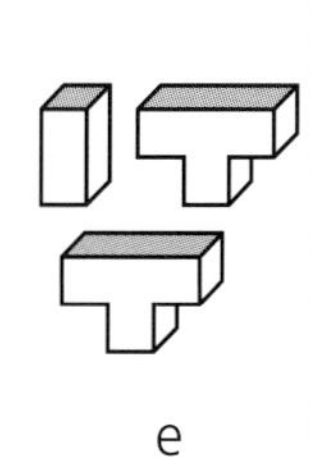

(a) b c d e

4

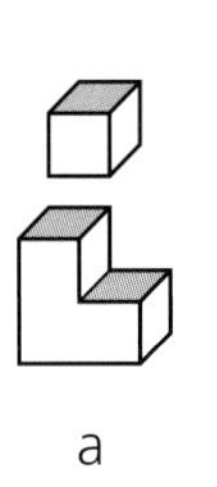
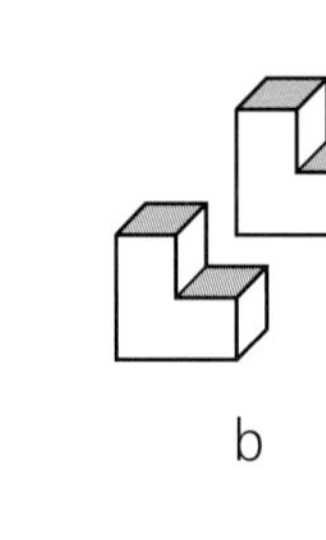
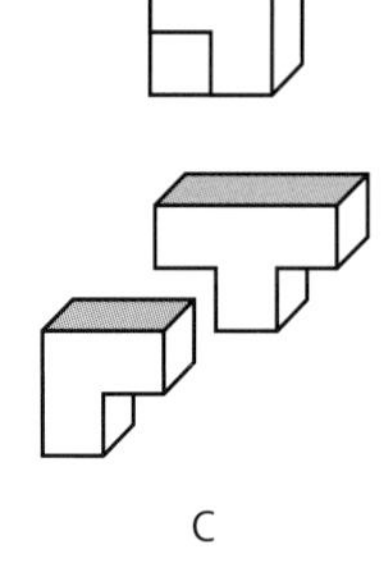
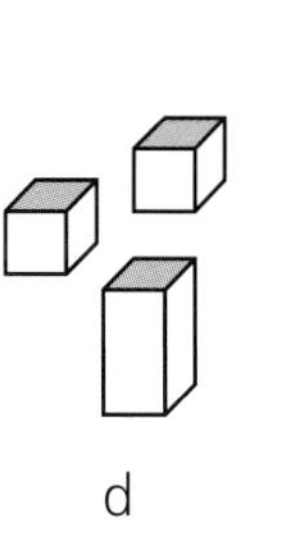
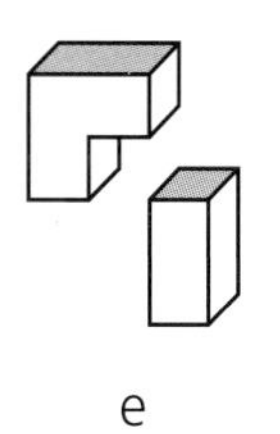

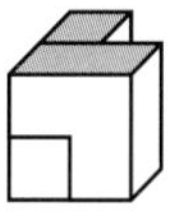

a b c d e

5

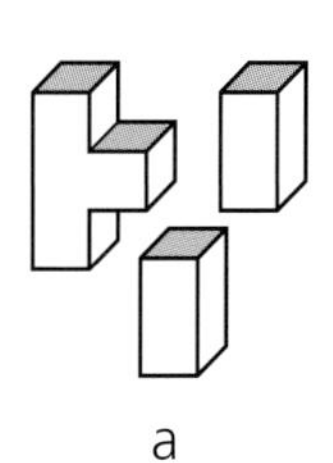
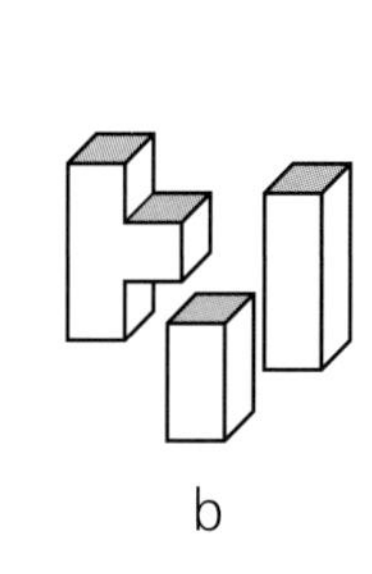
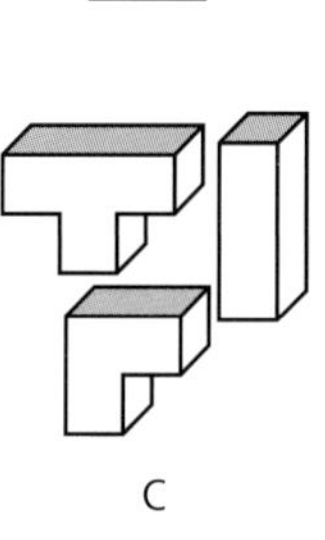
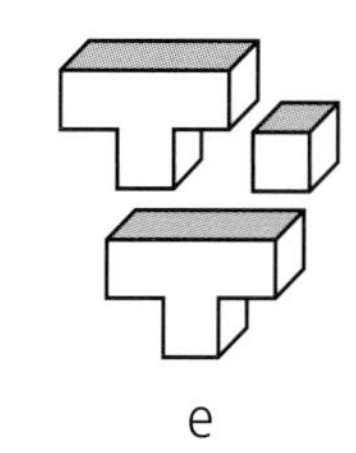

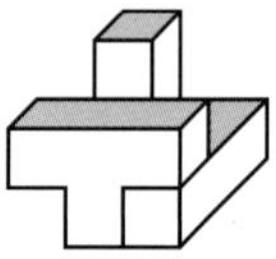

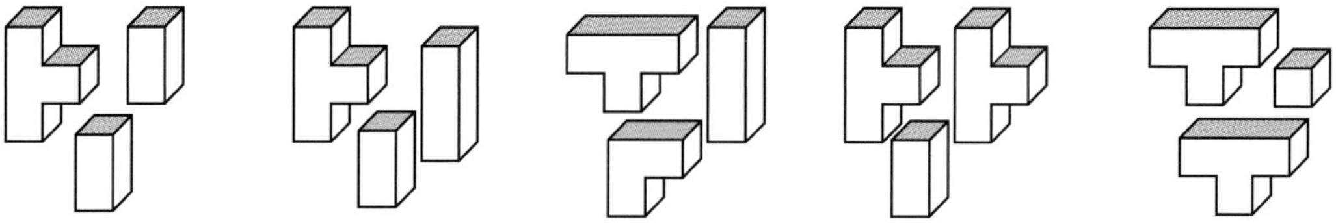

a b c d e

6

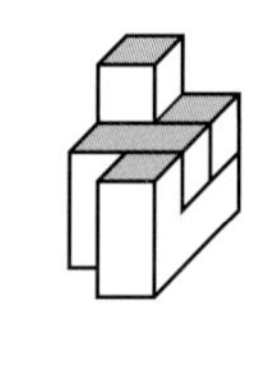

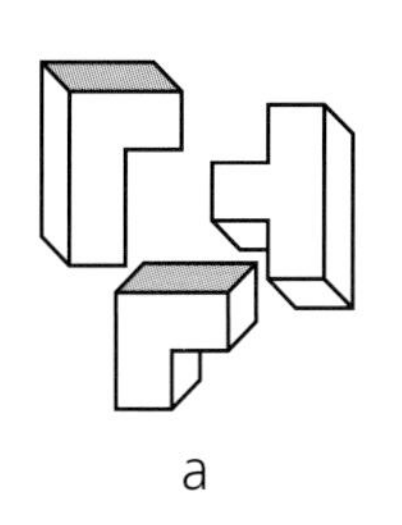
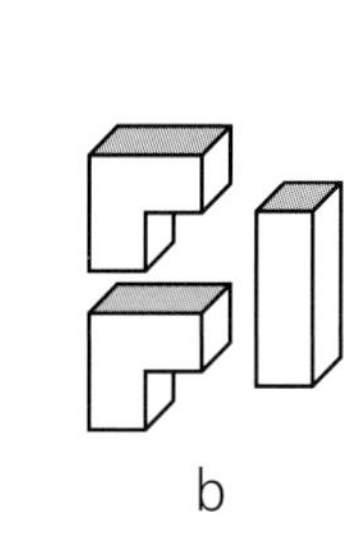
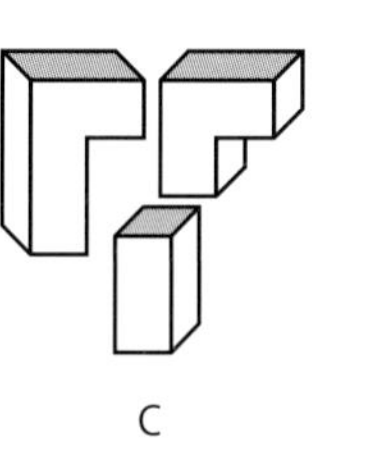
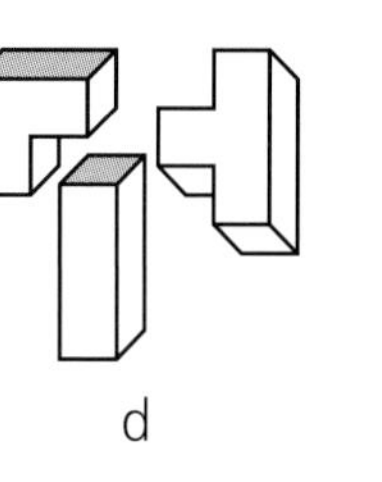
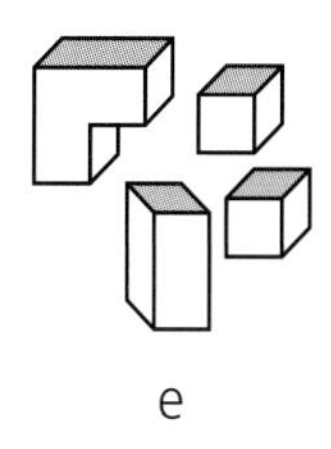

a b c d e

Test 3

Test time: 12:00

Complete this test in the time given above. Each question is worth one mark.

Look at the first two pictures and decide what they have in common. Then select the option from the five on the right that belongs in the same set. Circle the letter beneath the correct answer. Example:

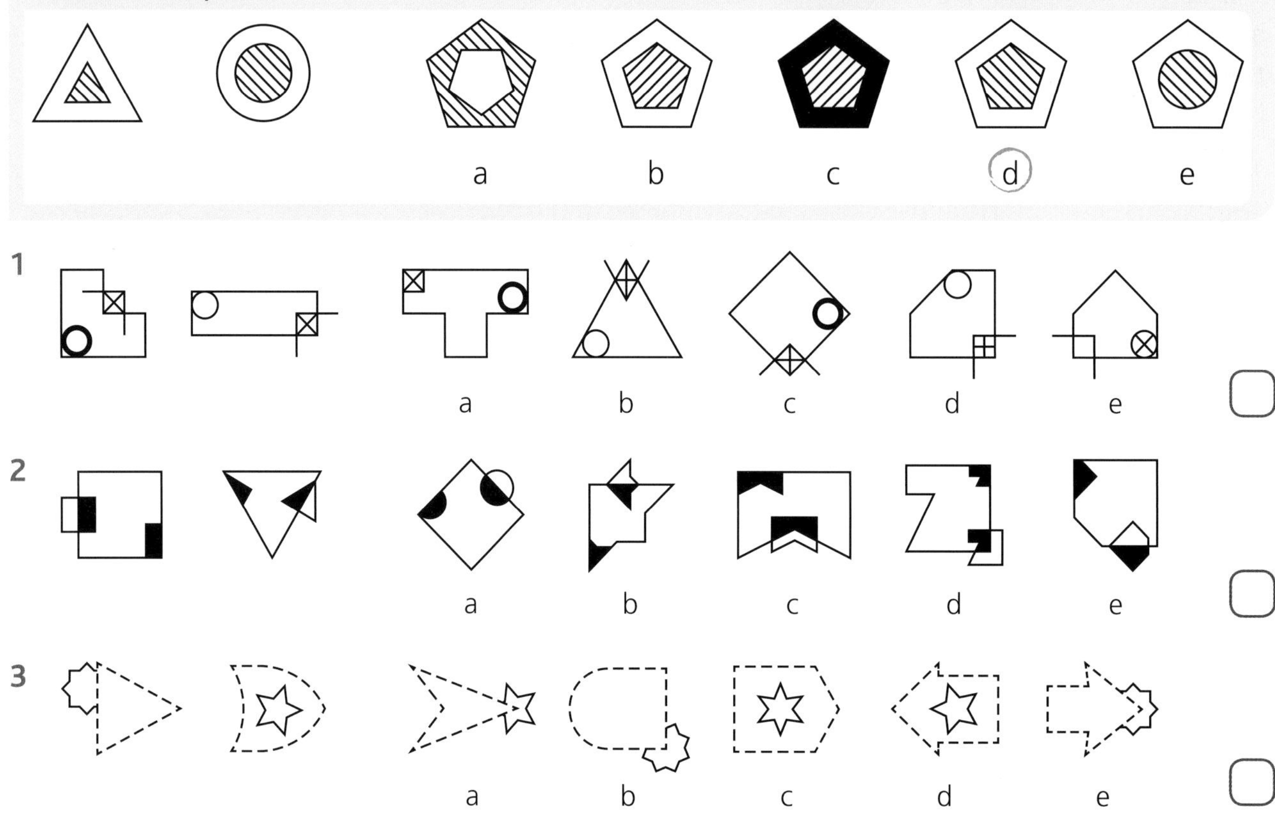

The shape on the left can be made up using three of the five smaller shapes on the right. Identify the **three** shapes needed and circle the letters beneath them. Example:

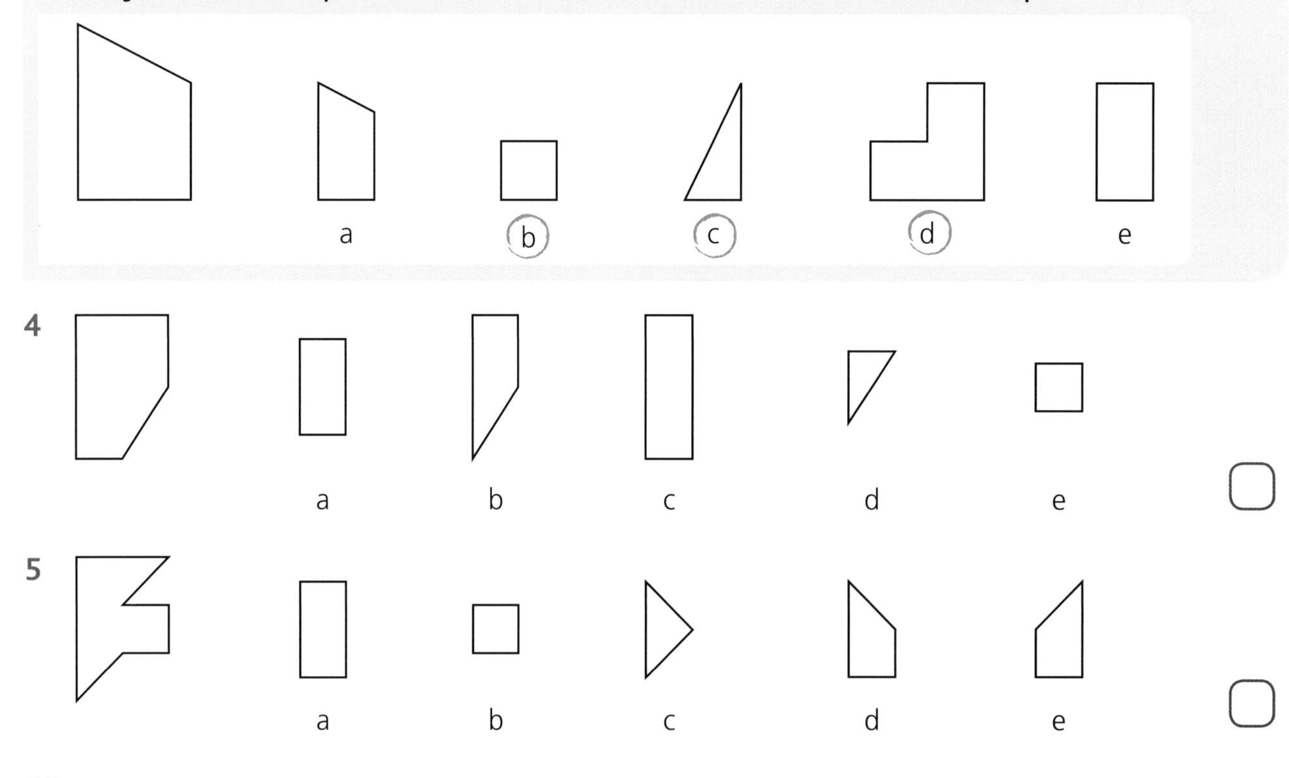

3 Links of position and direction

6

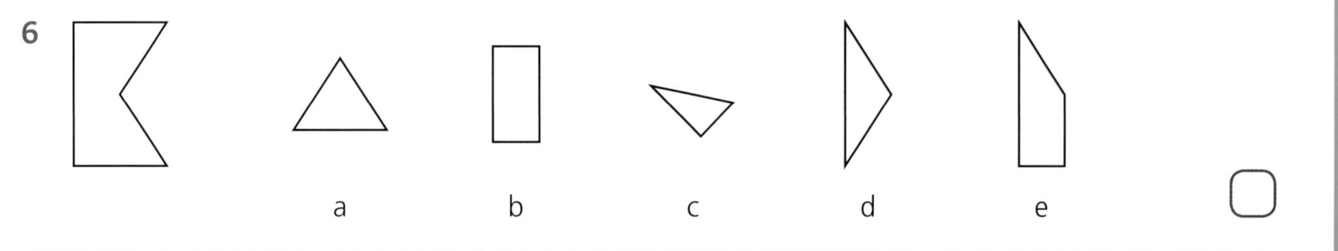

The square given at the beginning is folded in the way indicated by the arrows, and then holes are punched where shown on the final diagram. Identify the answer option that shows what the square would look like when it is unfolded. Circle the letter beneath the correct answer. Example:

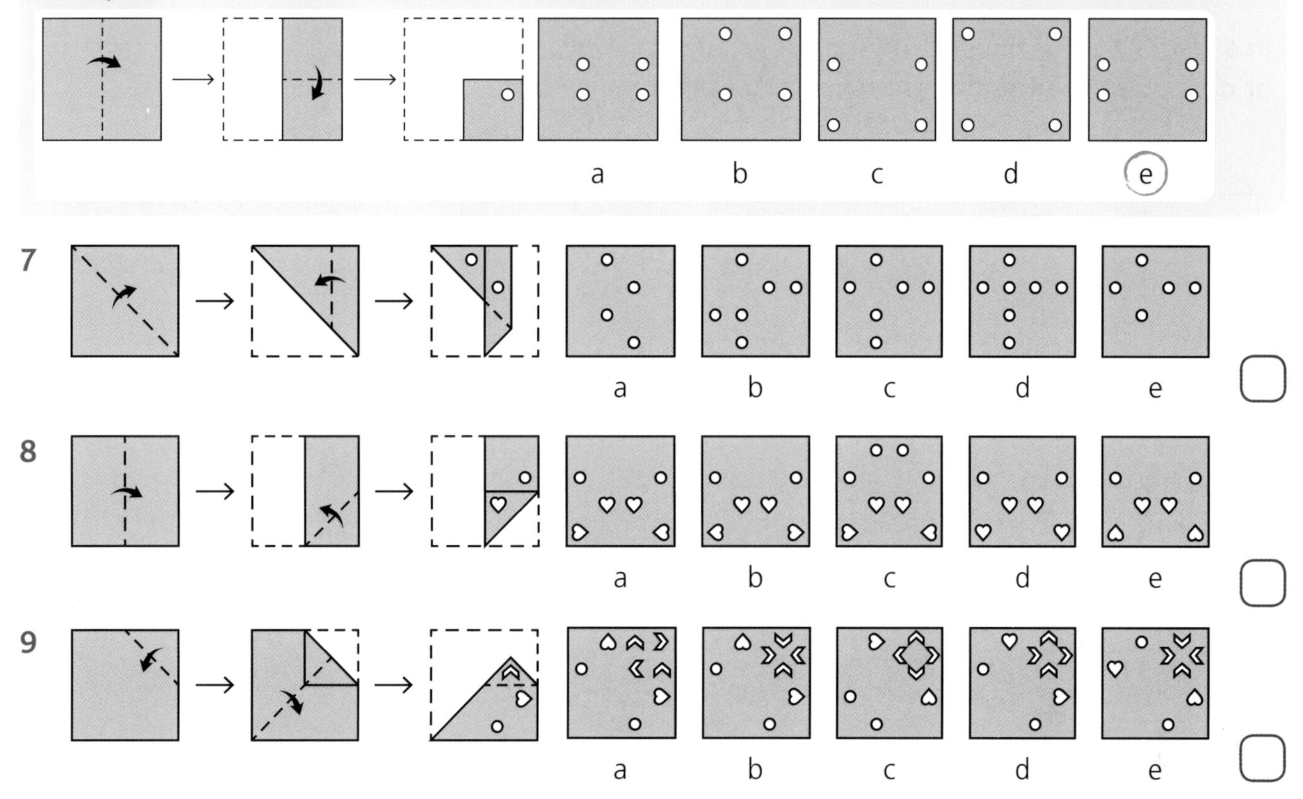

The picture on the left has been rotated clockwise by the number of degrees shown to give one of the pictures on the right. Circle the letter beneath the correct answer. Example:

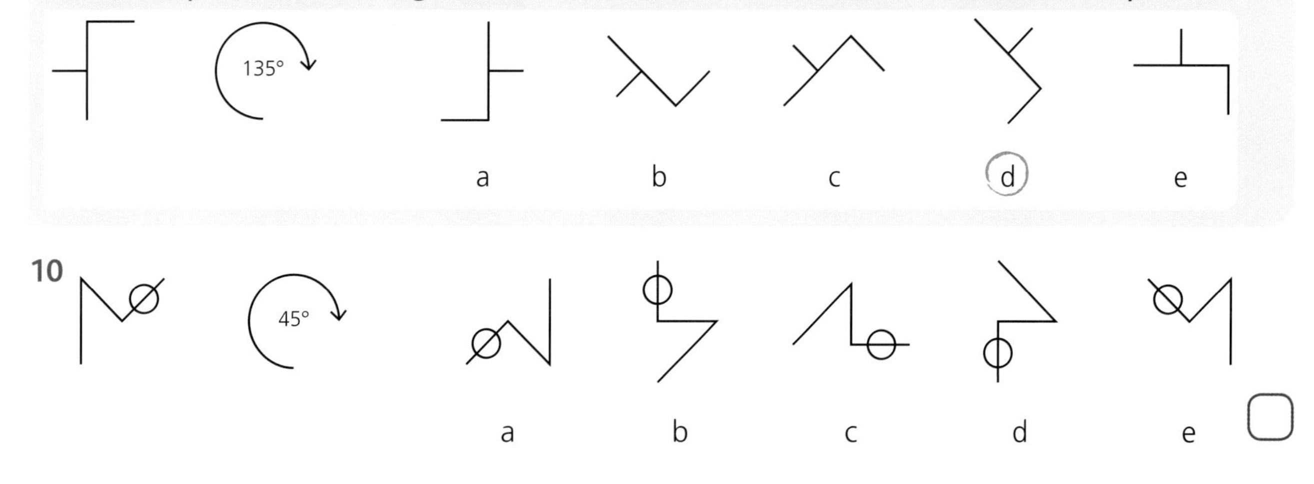

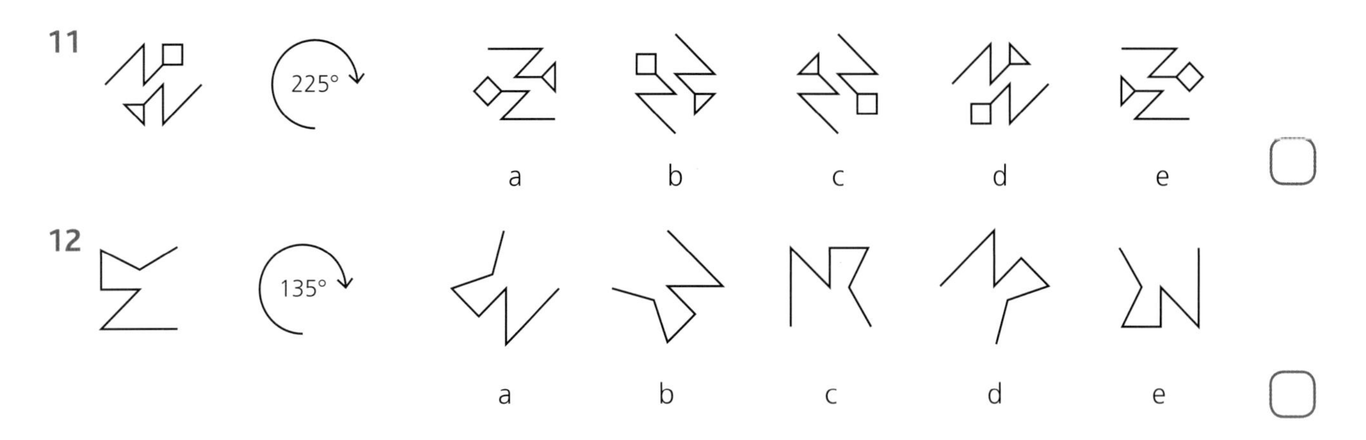

In questions 13 to 16, you will see a rotated version of one of the 3D diagrams shown (**a**, **b**, **c** or **d**). Circle the letter of the answer option that indicates the matching diagram above.

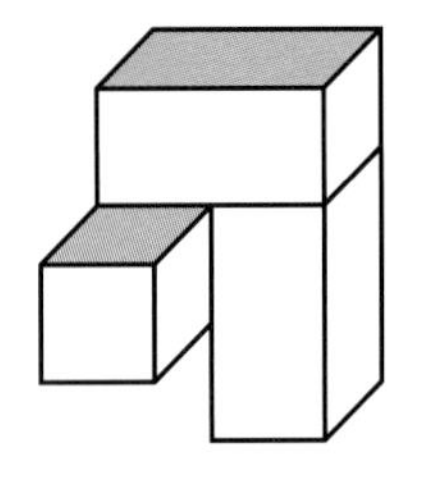

a

b

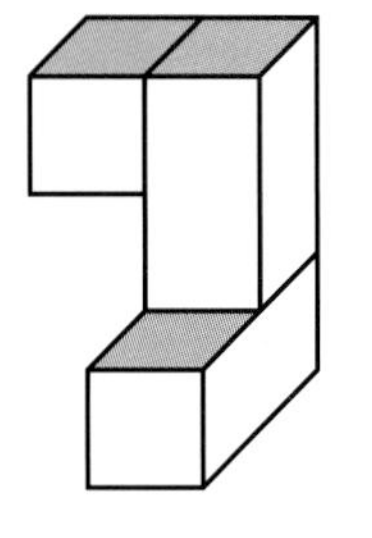

c

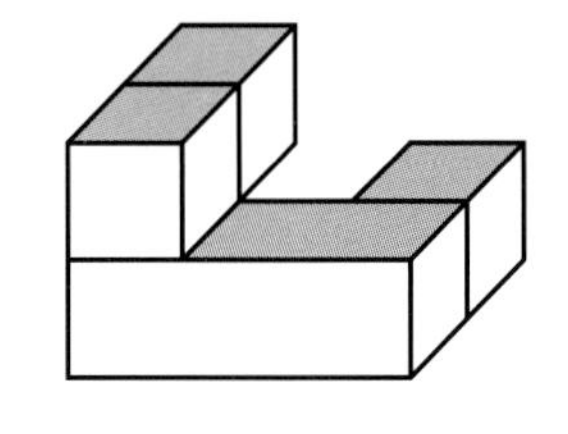

d

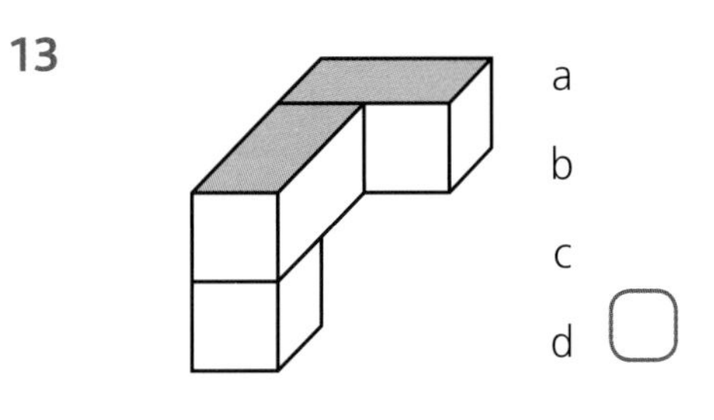

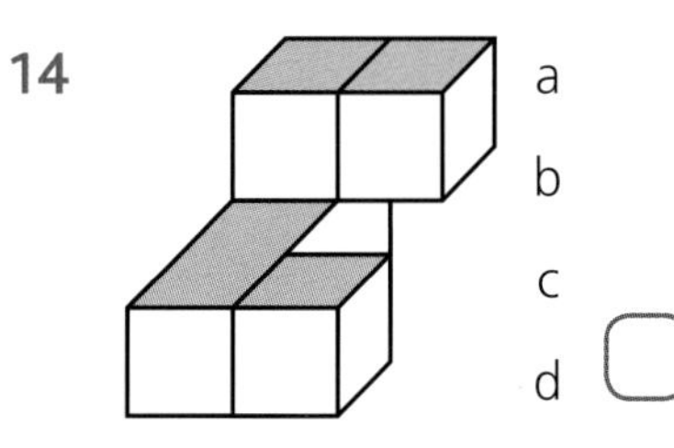

15

a

b

c

d

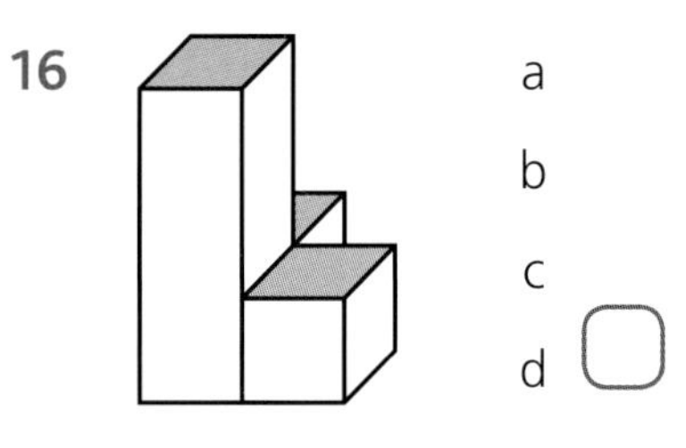

One group of separate blocks has been joined together to produce the pattern of blocks shown on the left. Some of the blocks may have been rotated. Circle the letter beneath the blocks that make up the pattern. Example:

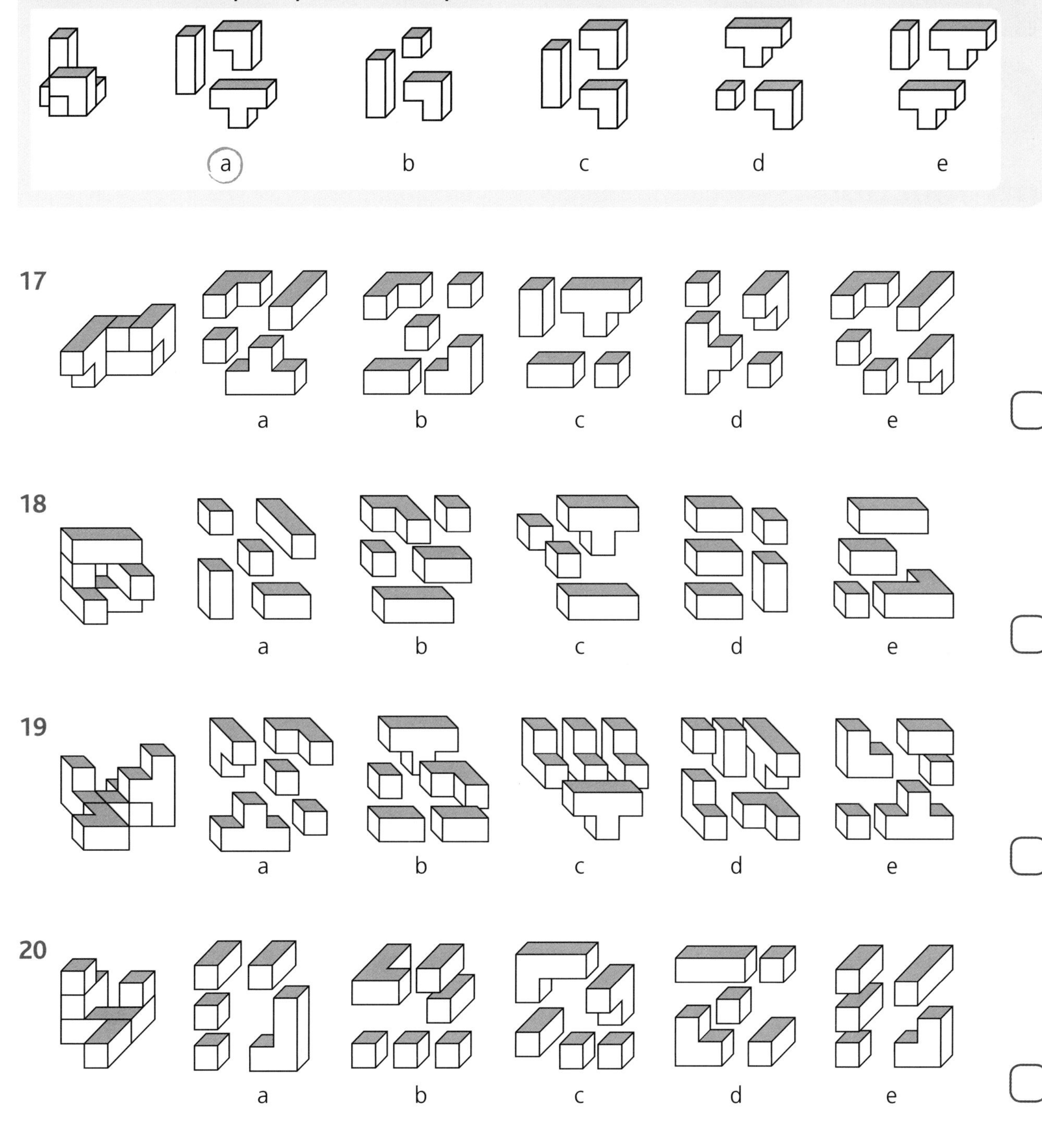

Record your score and time here and at the start of the book.

Part 3: Solving complex questions

4 Connections with codes

Introduction

In Part 2 you learned about more of the skills needed to solve Non-Verbal Reasoning questions. You are now going to look at how these skills work together in some of the more challenging question types.

Connections with codes questions use skills found in **algebra**, as described on page 20. You may be given one, two or three letters in a code to work out – the more letters, the more tricky the problems! In Maths the code letters stand for numbers, but in Non-Verbal Reasoning the letters stand for features in the pictures.

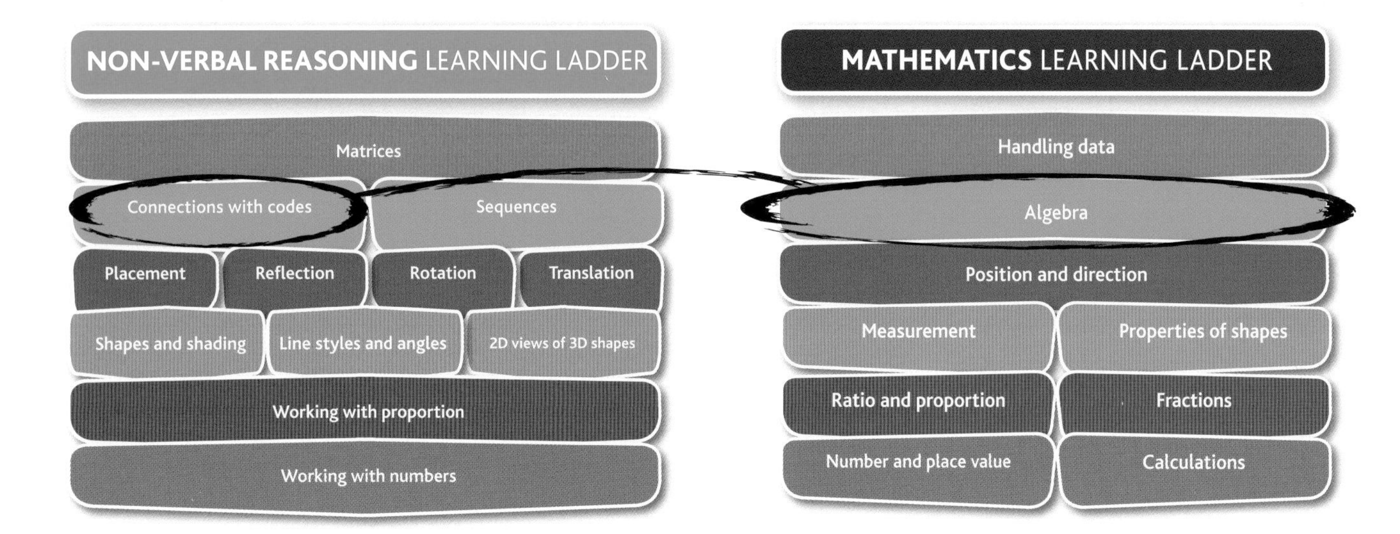

Have a look at the learning ladders again. You have already learned about the different features that can appear in Non-Verbal Reasoning questions in steps one to four and are now ready to tackle the challenging problems in step five.

You will see that algebra in Maths links across to the *Connections with codes* questions that are covered in this chapter, as well as the *Sequences* type that will be covered in Chapter 5.

Advice for parents

Children are in the early stages of working with algebra in school. The idea of applying a mathematical rule using letters can be hard to master, so children often have reservations about Non-Verbal Reasoning questions that appear to be using the same principles.

As there is no Maths involved in these questions, they are not always quite as daunting as they seem. They are, nonetheless, a step on from the earlier questions in this book where your child was expected to find a rule and apply it to a new picture: here they are asked to find up to three rules and then apply these rules to a new picture.

Avoid doing algebra practise to train your child in these questions as they will get more than enough of this at school! Instead, look for brainteaser questions (which are readily available on the internet) and the logic questions in the *Verbal Reasoning* books in this series as they will all provide good practice for solving code questions.

Thinking skills and games

The following games can help to develop the skills featured in this chapter.

Garden treasure trail

Create a treasure trail for a friend or parent, and then ask them to create one for you.

Draw a map of your garden, or find a map site showing a nearby area (such as a playing field) where you can safely leave some clues. Mark 'North' on your map – when the map is held upright. Also mark a starting point.

Next, create a code sheet for your treasure hunter that gives directions to a clue hidden at a mysterious location, for example:

'**1.** Six paces West **2.** Five paces North-East **3.** Beneath the fountain of Aquarius (a watering can!)'

Hide another clue here. This time use a code such as capital or coloured letters in a code sentence, for example:

'**2** roPes Are Commonly usEd by SailorS to tiE knots **AND** these kNots **W**ill last for **6** years, PArtiCularly whEn Sailing.'

When decoded, this clue reads: '2 paces SE and NW 6 paces.'

Make sure you leave a prize for your sleuth to treasure after the final clue so they will later do the same for you!

Codes using numbers and shapes

Skill definition: Solve *Code* questions that use numbers and shapes.

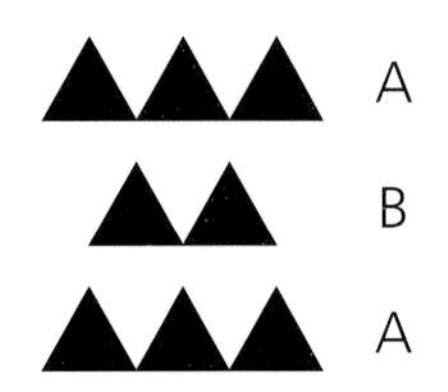

- *Code* questions can use any of the skills you have learned about in this book so far. On this learning spread we look at codes created using the areas of Maths you covered in Chapter 2:
  - numbers (shown by black triangles)
  - scale and proportion
  - 2D or 3D shapes and shading
  - line styles and angles.
- In the triangles above, **each letter** represents **one feature**. All *Code* questions work like this. Always **work left to right**.

Remember: You may come across distractions in *Code* questions. Do not expect every feature to count.

Method

Each letter represents an individual feature in the picture next to it. Work out which feature is represented by each letter. Apply the code to the picture in the box and circle the letter beneath the correct answer code.

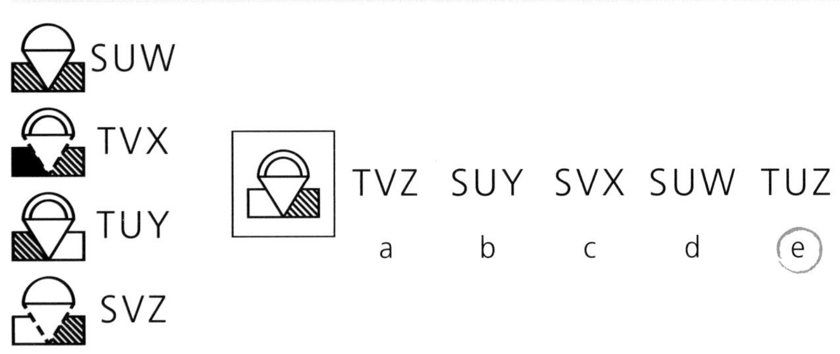

- Begin with the letters on the left. There are two Ss and two Ts, which tells you that:
  - the first and fourth (top and bottom) pictures have something in common
  - the second and third (middle two) pictures have something in common.
- Next, search for the common patterns in these pairs of pictures.
  - The first and fourth pictures both have a single semi-circle at the top.
  - The second and third pictures both have a double semi-circle at the top.

  So, the letter on the left stands for either a **single semi-circle** (**S**) or a **double semi-circle** (**T**).
- Now you can move on to the middle letter in each code.
  - The first and third pictures share a letter (U), as do the second and fourth pictures (V).
  - The triangle outline is solid in the first and third pictures and dashed in the second and fourth.

  This must mean that the middle letter in each code stands for the style of line used for the triangle: U for solid and V for dashed.
- You are nearly there, with only the final letter to work out:
  - All the letters in the final column are different (W, X, Y, Z) so there must be something that changes from picture to picture.
  - The only changing feature is the shading. This is different in each picture.
- So the third letter stands for the **shading** style. Your final task is to look at the picture in the box and **apply this code** to the image. Work from left to right to find the codes:
  - The picture has a double semi-circle (T), a triangle with a solid line (U) and diagonal shading in the right block (Z).
  - The code would therefore be TUZ so the answer must be option **e**.

Train

1 Work out the words made by these codes.

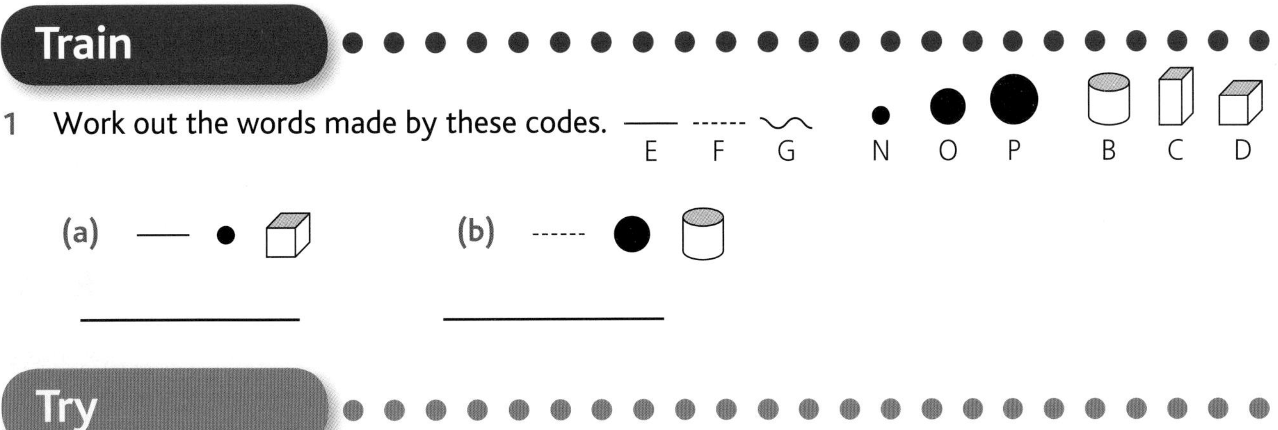

Try

Each letter represents an individual feature in the picture next to it. Work out which feature is represented by each letter. Apply the code to the picture in the box and circle the letter beneath the correct answer code.

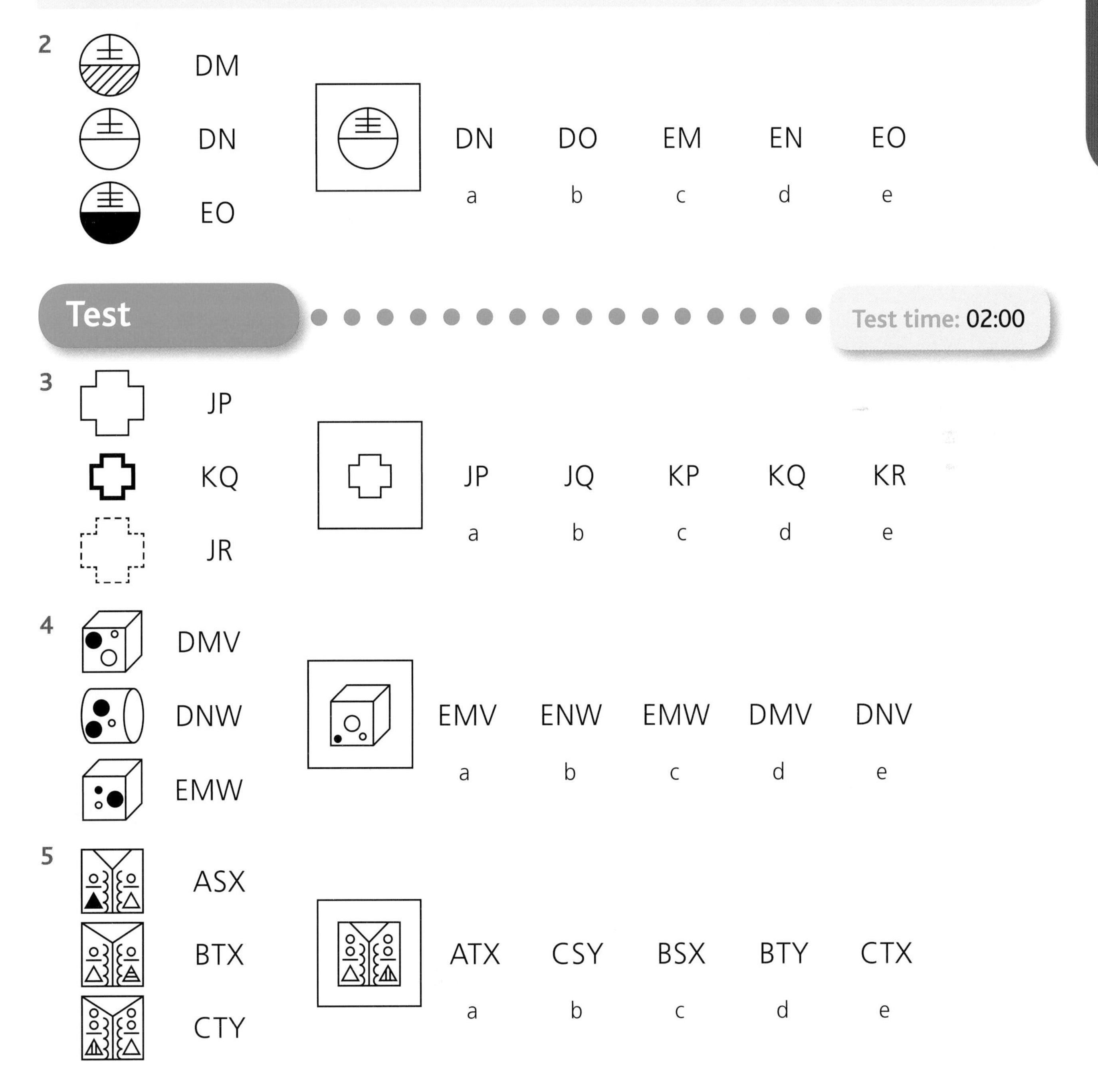

Codes using position and direction

Skill definition: Solve *Code* questions that use position and direction.

- On this learning spread you will be solving more *Code* questions but, this time, you will need to use the skills you learned in Chapter 3.
- You may see codes where:
  - shapes work in layers, or they touch
  - the direction or position of a shape is important
  - shapes are symmetrical or appear reflected.

Method

Each letter represents an individual feature in the picture next to it. Work out which feature is represented by each letter. Apply the code to the picture in the box and circle the letter beneath the correct answer code.

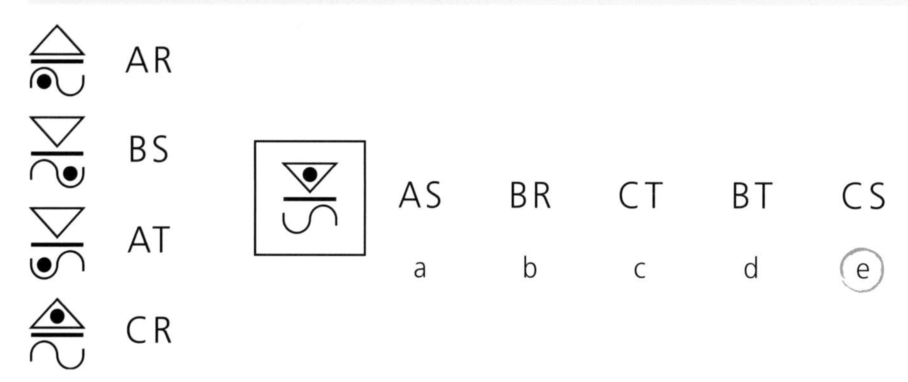

- Begin with the letter on the left as you did with the previous *Code* questions. This shows that the first and third pictures have something in common. The second and fourth pictures are different to each other and also to the first and third pictures.
- Look for a pattern linking the common pictures.
  - The position of the black circle is the only element that could link the first and third pictures.
  - The circle is on the right in the second picture and in the triangle in the fourth.

  This must mean that the first letter stands for the **position of the circle**: A for left, B for right and C in the triangle.
- Now move onto the second code letter.
  - The first and fourth pictures are the only ones with something in common as they have the same letter R. The triangle points up in both pictures.
  - To be sure the position of the triangle is the correct answer, check the other codes. The triangles point down in the second and third pictures but the codes are different. The curved lines are also different.
- As there is no clear pattern between the shapes, two **elements must be linked**.
  - As the circle is represented by the first code letter, the link is more likely to be between the triangle and the curved line.
  - R stands for the triangle pointing up with a curved line going up, then down; S stands for the triangle pointing down and the curved line going up, then down; T stands for the triangle pointing down and the curved line going down, then up.
- The code would therefore be CT so the answer must be option **c**.

If you think the first code contains a link between two elements, work through the other codes first and come back to it.

Train

1 Explain how the two pictures labelled **D** are linked.

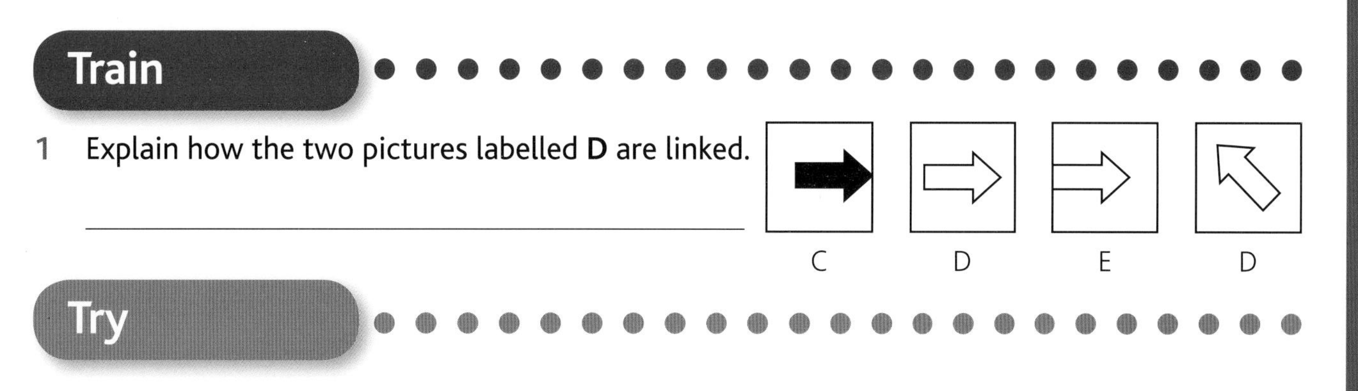

Try

Each letter represents an individual feature in the picture next to it. Work out which feature is represented by each letter. Apply the code to the picture in the box and circle the letter beneath the correct answer code.

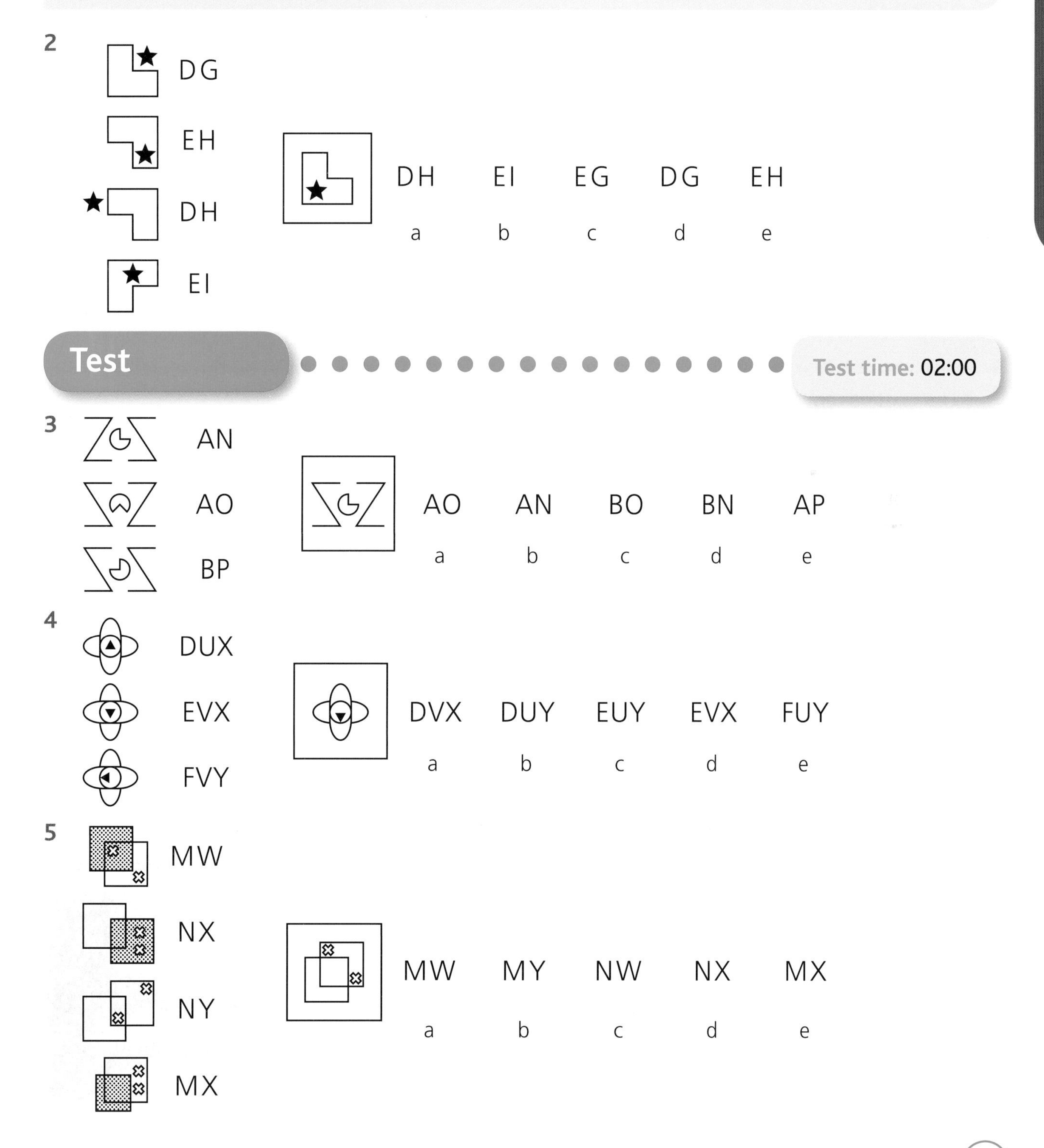

Codes combining different skills

Skill definition: Solve *Code* questions that use number, shape, position and direction and also questions where not all the codes are given.

- It is now time to combine all the skills you have learned in number, shape, position and direction to solve some more code questions. You may see questions with:
  - distractions
  - links between elements.
- You will also learn how to work out codes, even when there is no element representing the letter. The example below shows how to solve these questions.

Method

Each letter represents an individual feature in the picture next to it. Work out which feature is represented by each letter. Apply the code to the picture in the box and circle the letter beneath the correct answer code.

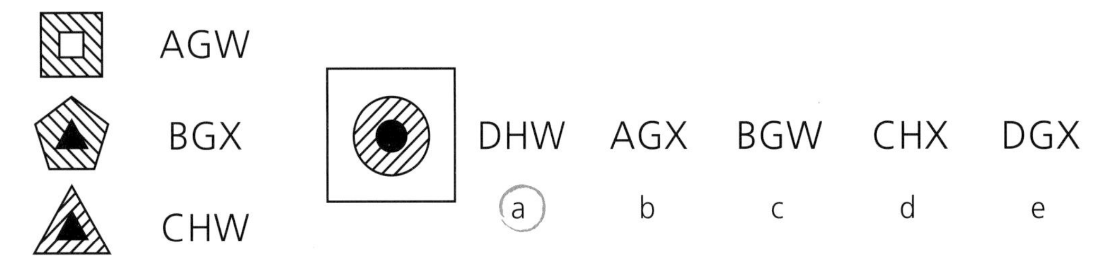

- Start with the first letter: you are looking for a feature that is different in all three pictures.
  - The outer shapes are all different so this must be the rule for the first code.
- For the middle code letter, look for a connection between the top two pictures (both letter G).
  - The diagonal shading matches in the top two pictures but changes in the bottom picture.
  - This is a clear connection, so we know the second code letter stands for the **direction of shading in the outer shape**.
- Next look for patterns to solve the last code. You know there is a connection between the top and bottom pictures as both have the letter W, although it is not clear what this might be.
  - The inner and outer shapes in both the top and bottom pictures match, but they are different in the middle picture.
  - So this third code stands for whether or not the picture has two matching shapes.
- Finally, look at the picture in the box to work out its code.
  - As soon as you look at it you will see a problem: we know the first letter stands for the outer shape, but there are no circles in the original set of pictures. This must mean that there is an extra code letter in the answers.
  - The code letters for the existing shapes are A, B, C, so this extra letter will probably follow on in the alphabet.
  - Looking at the answer options, **a** and **e** have a letter D.
- Work out the remaining two codes (H for direction of shading and W for identical shapes), so the answer must be option **a** (DHW).

You may also see *Code* questions where the letters are in different positions. These all work in the same way, so do not be put off.

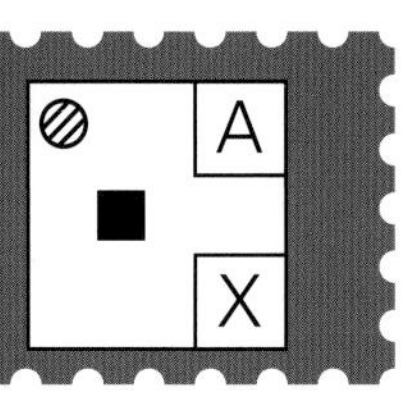

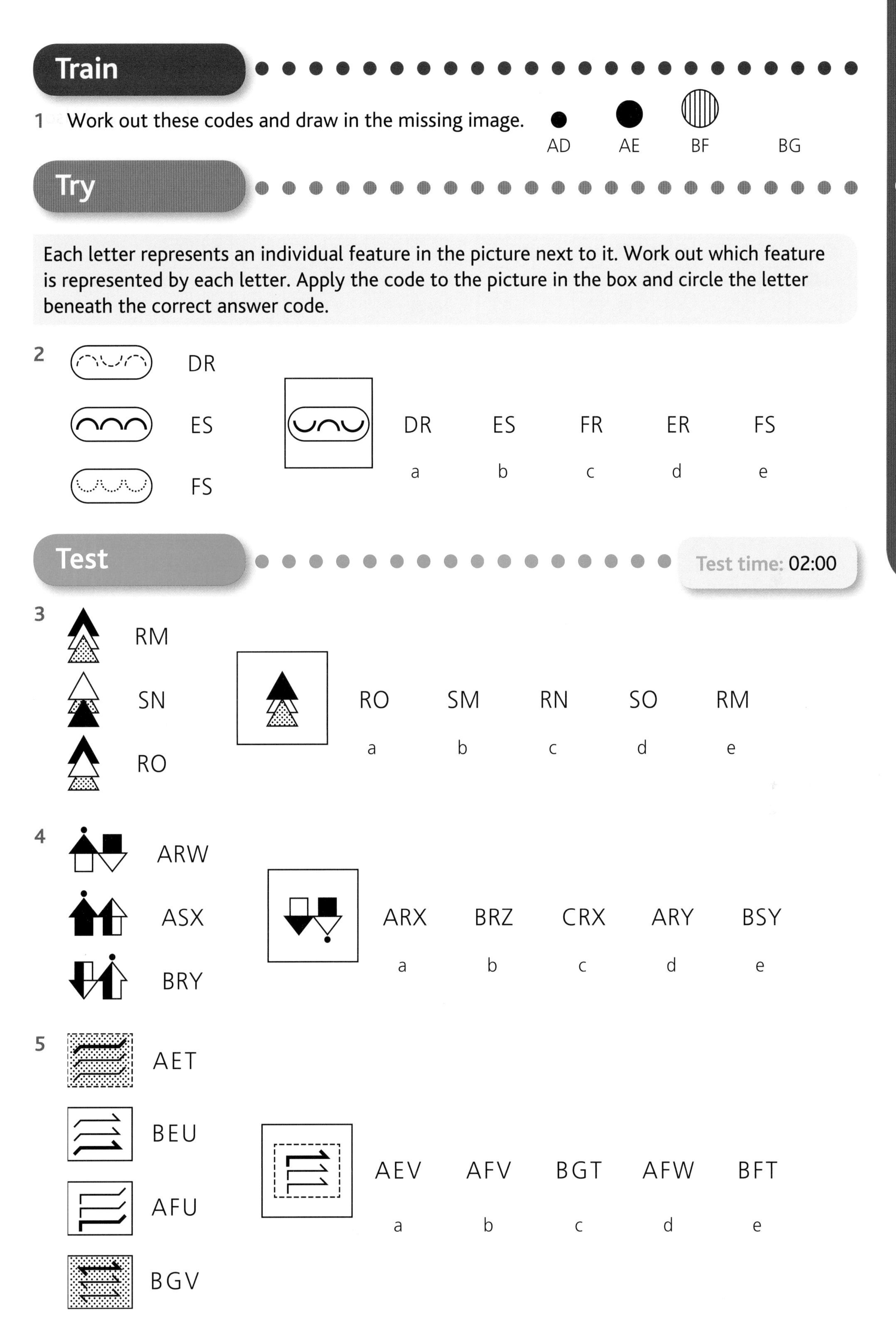

Train

1 Work out these codes and draw in the missing image.

AD AE BF BG

Try

Each letter represents an individual feature in the picture next to it. Work out which feature is represented by each letter. Apply the code to the picture in the box and circle the letter beneath the correct answer code.

2 DR ES FS

DR	ES	FR	ER	FS
a	b	c	d	e

Test

Test time: 02:00

3 RM SN RO

RO	SM	RN	SO	RM
a	b	c	d	e

4 ARW ASX BRY

ARX	BRZ	CRX	ARY	BSY
a	b	c	d	e

5 AET BEU AFU BGV

AEV	AFV	BGT	AFW	BFT
a	b	c	d	e

Test 4

Test time: 05:00

Complete this test in the time given above. Each question is worth one mark.

The two letters in the small boxes at the right of each large box represent a feature of the shapes in the box. Work out which feature is represented by each letter and apply the code to the box with the dashed lines. Circle the letter beneath the correct answer code. Example:

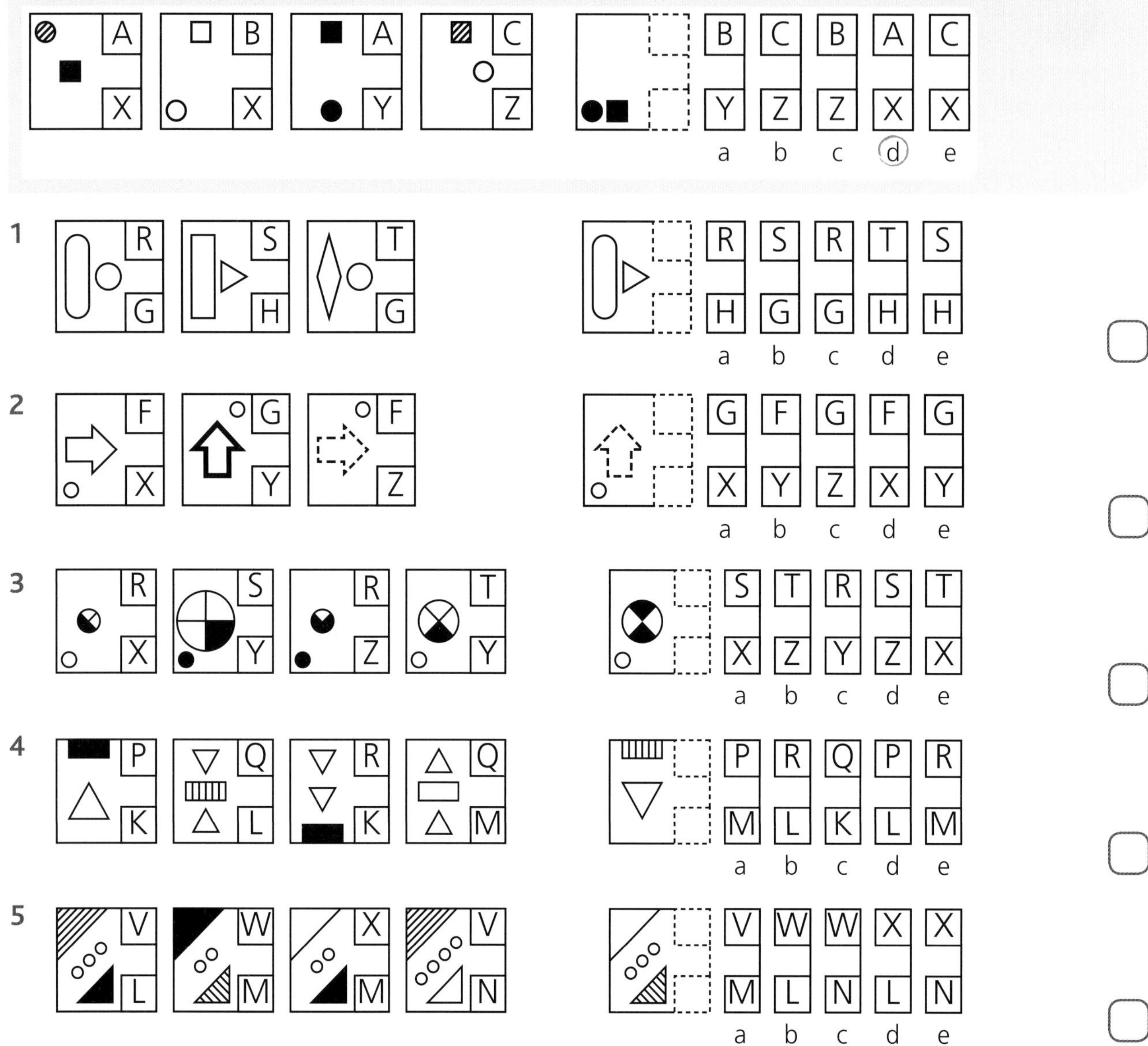

Each letter represents an individual feature in the picture next to it. Work out which feature is represented by each letter. Apply the code to the picture in the box and circle the letter beneath the correct answer code. Example:

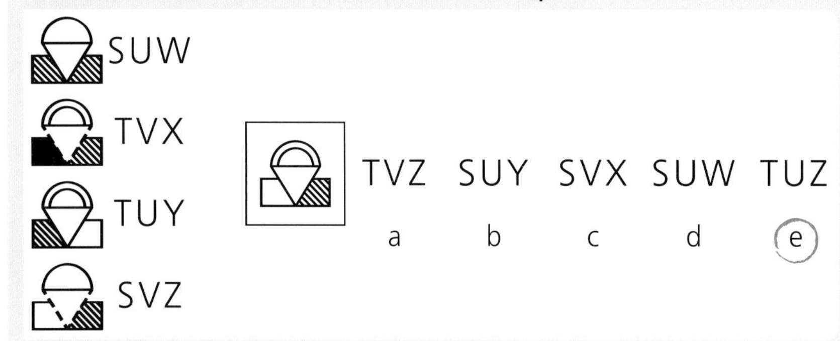

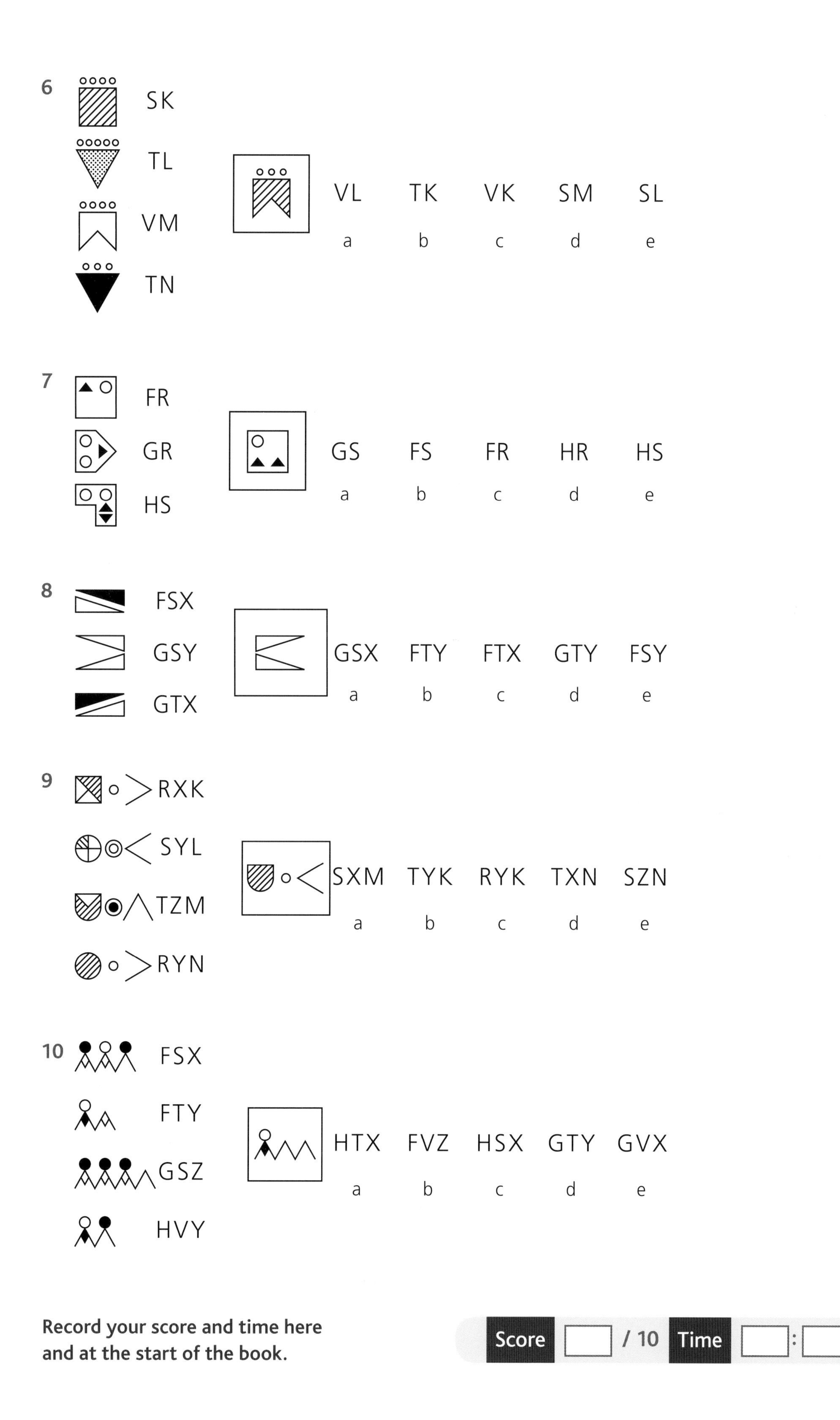
6
SK
TL
VM
TN
VL TK VK SM SL
a b c d e
7
FR
GR
HS
GS FS FR HR HS
a b c d e
8
FSX
GSY
GTX
GSX FTY FTX GTY FSY
a b c d e
9
RXK
SYL
TZM
RYN
SXM TYK RYK TXN SZN
a b c d e
10
FSX
FTY
GSZ
HVY
HTX FVZ HSX GTY GVX
a b c d e
Record your score and time here
and at the start of the book.
Score
/ 10
Time

5 Sequences

Introduction

You have had plenty of practice spotting features that match and change between pictures in the previous chapters. Now you are going to do exactly the same thing with a row of pictures.

Just like sequences in Maths, you are looking for a pattern that changes from one step to the next. The difference is that the pattern is between pictures rather than numbers.

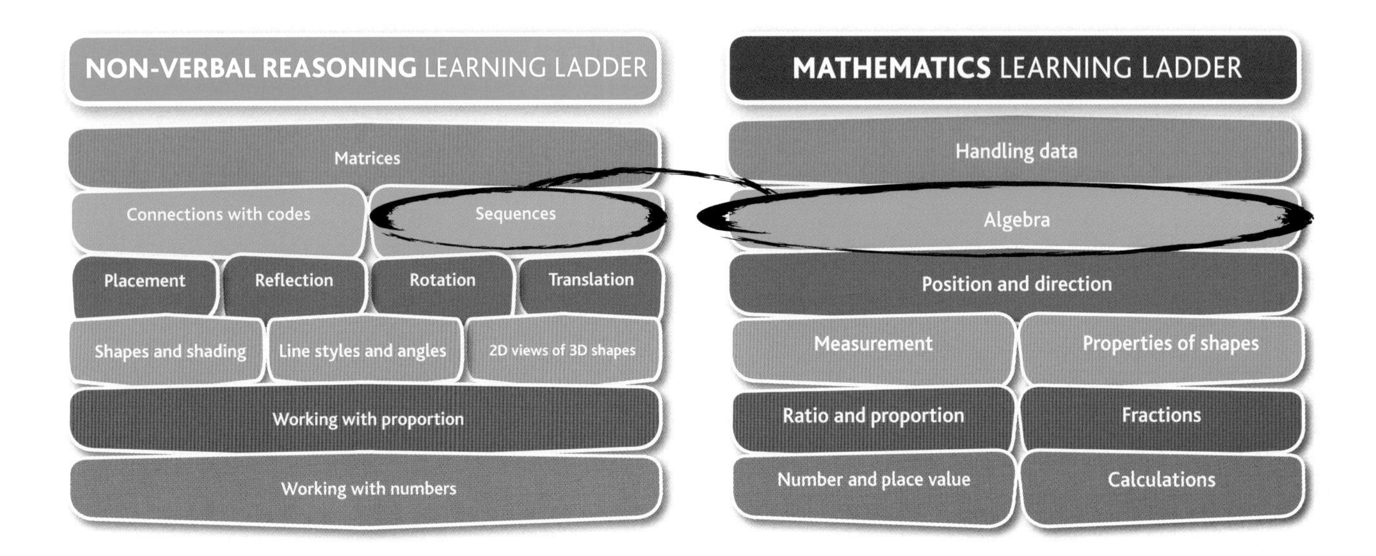

You will see on the learning ladders that algebra in Maths not only links to *Codes* but also to *Sequences*. You are almost on the final step of the ladder, so get ready for some tricky questions to prepare you for stepping right up to the top in the next chapter!

Advice for parents

Sequence questions in Maths involve looking for rules in a sequence of numbers, such as adding 5, multiplying by 2 and so on. Geometric sequences are now also included in Maths teaching and involve the same logic skills where you are expected to find patterns from one step to the next.

Non-Verbal Reasoning sequences work in a similar way and build on the *Matching features* and *Applying changes* questions covered earlier on this book (mostly in Chapter 2). These questions looked at elements that stay the same or change between pictures. *Sequences* questions expect your child to spot these features in a series of pictures.

The Maths skills reviewed earlier in this book – looking at simple number calculations and properties of shapes – are important in *Sequences* questions and so revision in either of these areas is helpful to raise your child's awareness of the kind of tricks examiners employ in *Sequences*.

Thinking skills and games

The following games can help to develop the skills featured in this chapter.

Wrap it up!

A fun way to explore how patterns build into sequences is to create your own using small objects such as leaves and flowers.

Explore using similar shapes, sizes, contrasting colours and rotations. Photograph the pattern and use the image to create wrapping paper and tags. Rotate the sequence and see how it works when you piece it together to form a continuous pattern, as in the example above.

Number snap

You will need a standard deck of playing cards as well as the specified number cards to play this game for two to four people.

Take out the picture cards and jokers from the deck of cards. Then write or print the following rule cards: +1, +2, +3, –1, –2, –3

1. The dealer deals out the whole deck of cards between the players. Everybody should keep their cards in a pile, face down, in front of them.
2. Before beginning, the dealer shuffles the number cards and turns the top one face up. Whatever is shown on the card, for example '–1', is the rule for this game.
3. Play clockwise from the dealer, with each player turning up the top card from his or her pile and placing it on a pile in the centre. As soon as somebody spots a card that follows the rule – for example, if the rule is '–1' then a 5 following a 6 would be correct – that player should shout 'Snap!'
4. Anybody calling 'Snap!' at the wrong time has to give their top card to the person on their left.
5. Carry on playing by turning over the next rule card to set a new rule.
6. The winner is the person with all the cards at the end.

Sequences using numbers and shapes

Skill definition: Solve *Sequences* questions that use numbers and shapes.

- *Sequences* questions can use any of the skills you have learned about in this book so far. The challenge is to work out what goes in the **missing box** in a question, so you need to follow the patterns carefully.
- In this learning spread you will find out about sequences created using the areas of Maths explained in Chapter 2: that is, **numbers**, **scale** and **proportion**, **2D** and **3D shapes** and **shading** as well as **line styles** and **angles**.
- Sequences can:
  - repeat
  - reflect
  - **use numbers** which get bigger or smaller
  - **work in steps** where something changes each time.

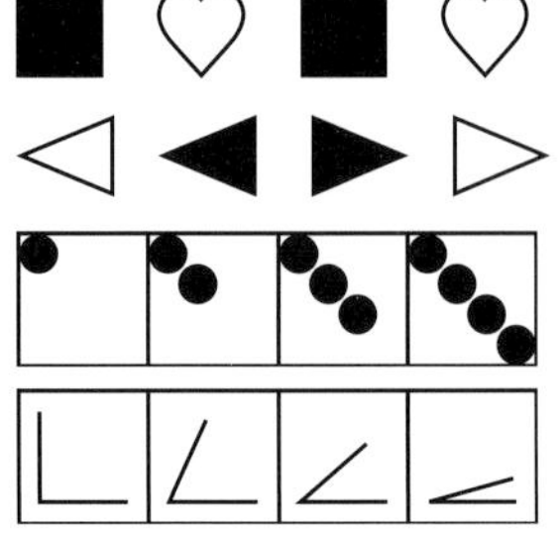

- Many sequences have **two or more things happening**, as shown in the example below.

Method

The boxes on the left show a pattern that is arranged in a sequence. Choose the answer option that completes the sequence when inserted in the blank box. Circle the letter beneath the correct answer.

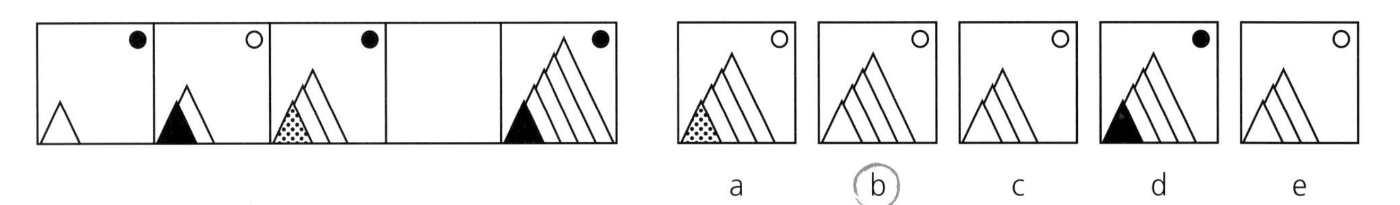

- The trick with these questions is to look at **one feature at a time** to try and find out what is going on.
  - The circles and the bottom triangle are sometimes shaded, but the shading does not seem to match.
  - The triangles change in number but not shape.
  - The circles change in shading but not size or position.
- Because the shading **does not match** in the circle and triangle, they are probably following a **different pattern**. So, there are **three patterns** you need to look at: the shading of the circle, the shading of the triangle and the number of triangles.
- Now try to follow each pattern. Start with the triangles.
  - One extra triangle is added each time in the first three boxes, following a **simple number pattern** (one in the first box, two in the second and three in the third). If the pattern continues, there should be four triangles in the missing box and five in the final box (which there are).
  - The extra triangle is always larger than the previous one.
  - So, there should be four triangles of different sizes in the box meaning **c** and **e** can be rejected.
- Now move on to the shading.
  - The circle changes from black to white and then back to black again. If this pattern **repeats**, then the missing box would be white and the final box would have a black circle (which it does).

- The bottom triangle changes from white to black to dotted. If this pattern **repeats every three boxes**, the triangle in the missing box would be white and in the final box black (which it is).

- If this is correct, you will be able to find an answer option with four triangles, a white circle and a white triangle at the bottom. This matches option **b** which must be the correct answer.

Train

1 Add three more pictures to this sequence where something is added each time – use your imagination!

Try

The boxes on the left show a pattern that is arranged in a sequence. Choose the answer option that completes the sequence when inserted in the blank box. Circle the letter beneath the correct answer.

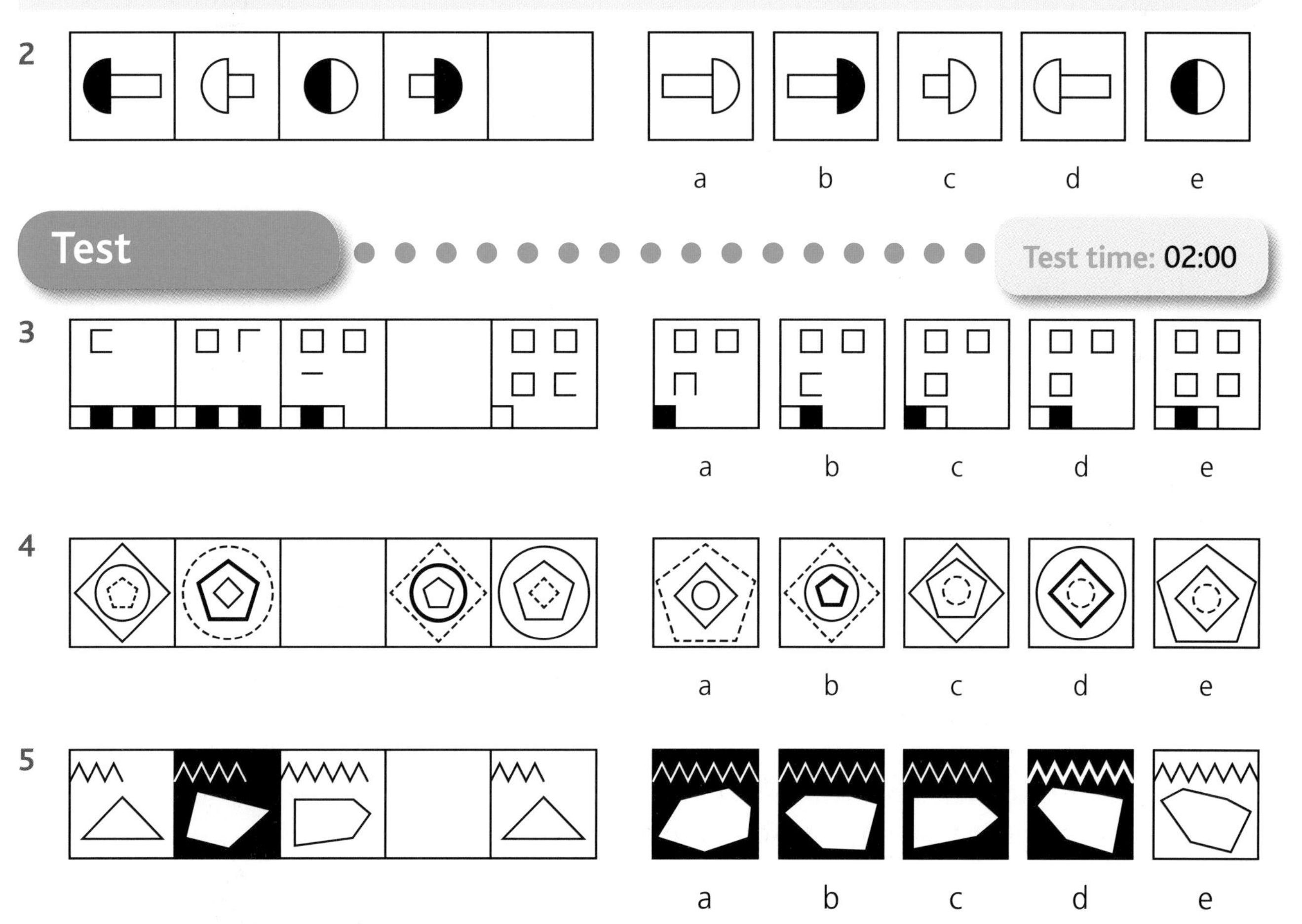

2 a b c d e

Test

Test time: 02:00

3 a b c d e

4 a b c d e

5 a b c d e

Sequences using position and direction

Skill definition: Solve *Sequences* questions that use position and direction.

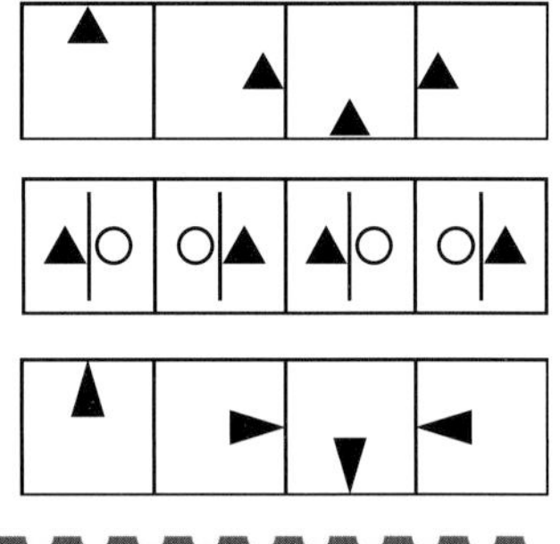

- On this learning spread you will be solving more *Sequences* questions but, this time, you will need to use the skills you learned in Chapter 3: shapes working in **layers**, patterns using **direction** and **rotation**, and shapes that are **reflected**.
- Many sequences involve position and rotation where shapes:
  - move around a box (top picture on the right)
  - move in relation to each other (middle picture)
  - rotate within a box (bottom picture on the right).
- You will also see **shading** that **rotates** and **moves** within shapes, as in the example below.

You are unlikely to see distractors in *Sequences* questions, so expect each feature to hold a clue to the correct answer.

Method

The boxes on the left show a pattern that is arranged in a sequence. Choose the answer option that completes the sequence when inserted in the blank box. Circle the letter beneath the correct answer.

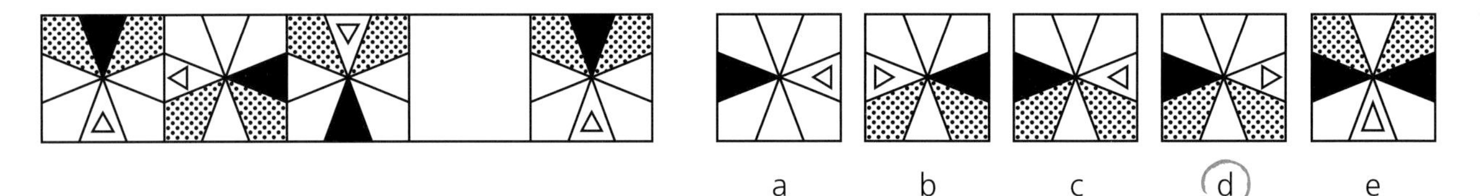

- Begin by looking closely at the sequence and **note the features that change**. It is important to do this so that you do not miss anything important when checking the answer options.
  - The dotted background changes position but not number.
  - The black shading moves around the 'windmill' shape.
  - The small triangle moves around the 'windmill' shape.
- So, now you know there are **three features** to look at. Work through them one at a time and start rejecting some answer options.
  - The dotted background moves from top to bottom from one square to the next so is a **repeating** pattern **every two squares**, so the shading will be at the bottom – reject answer options **a** and **e**.
  - The black triangle on the 'windmill' moves one shape clockwise from one square to the next, therefore **repeats every four squares**, so the left-hand triangle will be black – reject **b**.
  - The small triangle moves one 'windmill' shape clockwise and **rotates anticlockwise**, so it will be in the right-hand large triangle and point to the right – reject **c**.
- Check all the answer options: option **d** has dotted shading at the bottom, a black triangle on the left and a small triangle on the right **pointing to the right**. This must be the correct answer.

Remember: Look very carefully at the rotation of objects. This often holds the key to the answer.

Train

1 Green and Blue frog are friends. Green frog can only hop half a side clockwise while Blue frog can only hop half a side anticlockwise. Add boxes and draw the sequence to show how many hops it takes the frogs to meet.

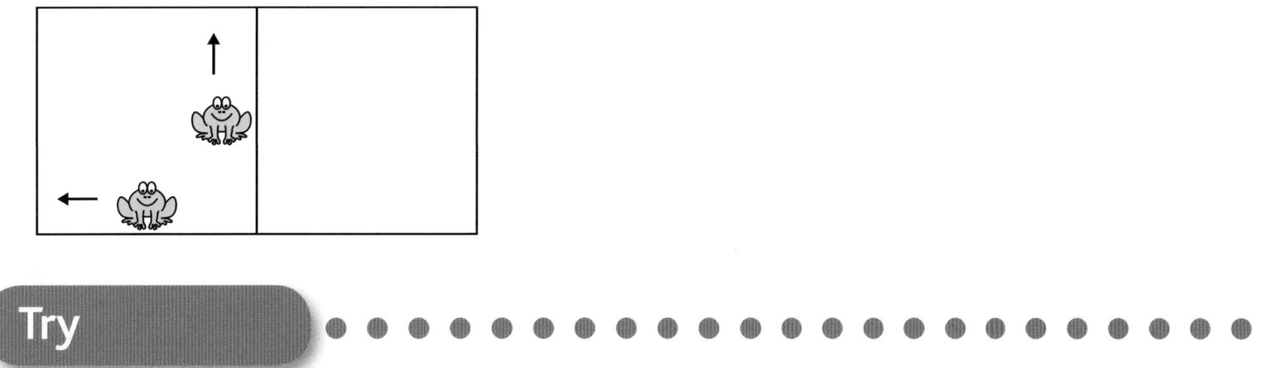

Try

The boxes on the left show a pattern that is arranged in a sequence. Choose the answer option that completes the sequence when inserted in the blank box. Circle the letter beneath the correct answer.

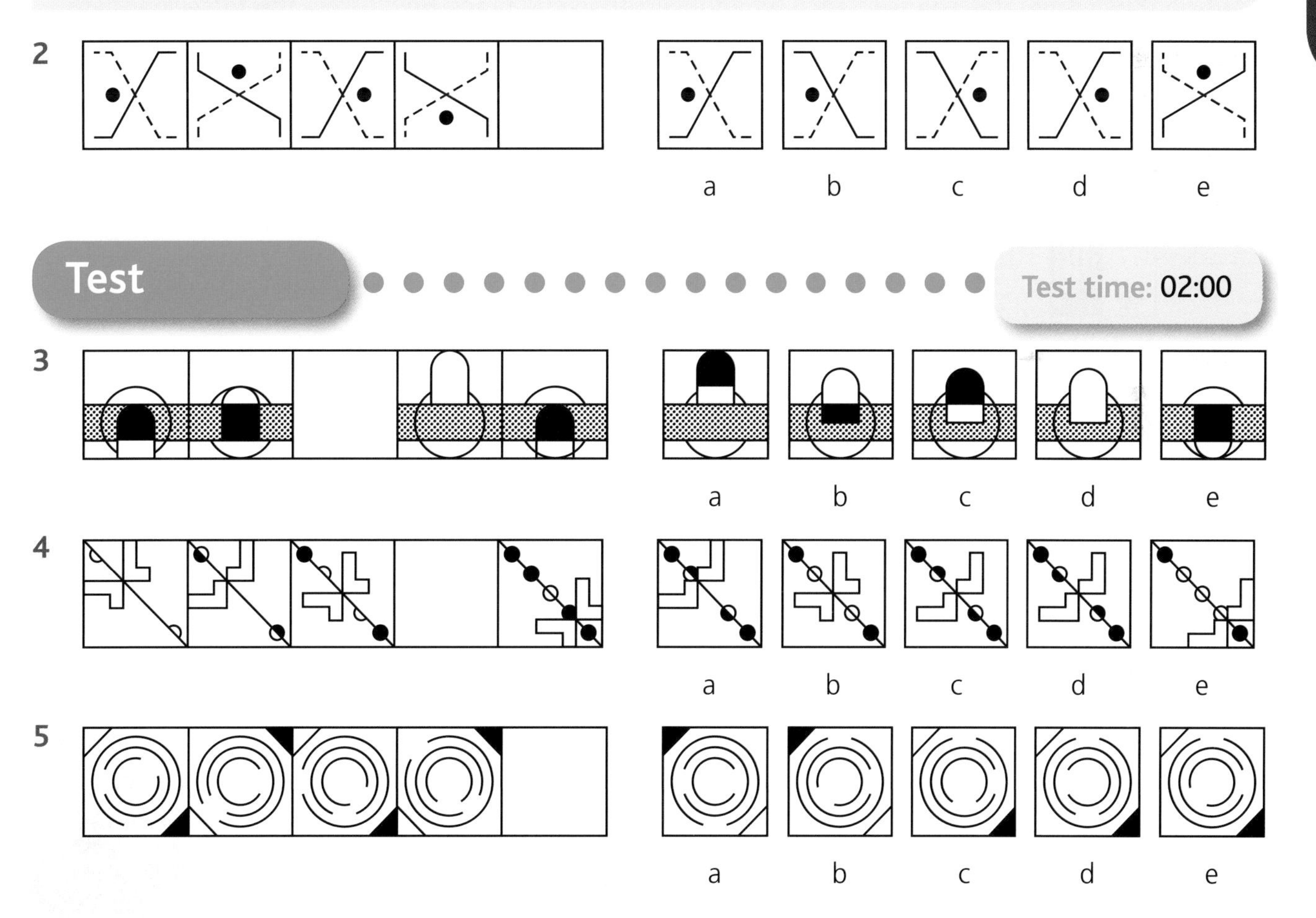

Sequences combining different skills

Skill definition: Solve *Sequences* questions arranged in different ways that use number, shape, position and direction.

- It is now time to combine all the skills you have learned for number, shape, position and direction to solve some more *Sequences* questions.
- It is possible for **any box** in the sequence to be **missing**. If the first box is missing, you just have to remember to work out how the sequence will work backwards, for example:
  - if a number sequence **adds** one, remember to **subtract** one
  - if a rotation works **clockwise**, remember to look for an answer that is anti**clockwise**.
  - Sequences vary in the number of steps – the more steps you have, the more clues you have to find the answer. Also look out for **split** sequences like the one below.

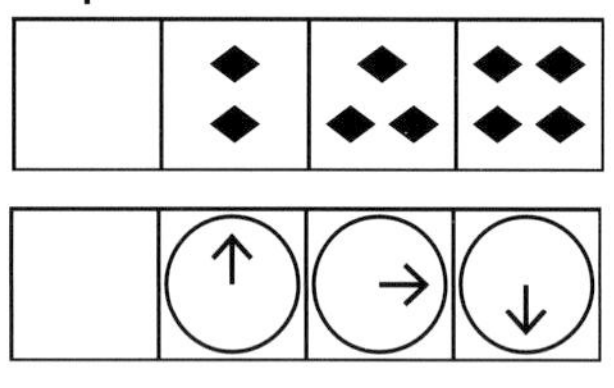

Method

The boxes on the left show a pattern that is arranged in a sequence. Choose the answer option that completes the sequence when inserted in the blank box. Circle the letter beneath the correct answer.

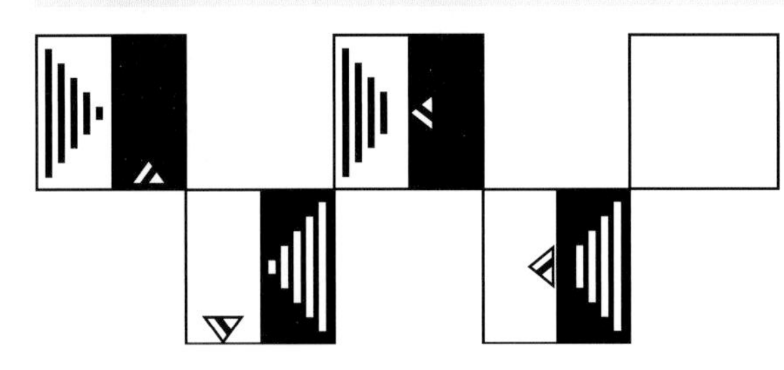

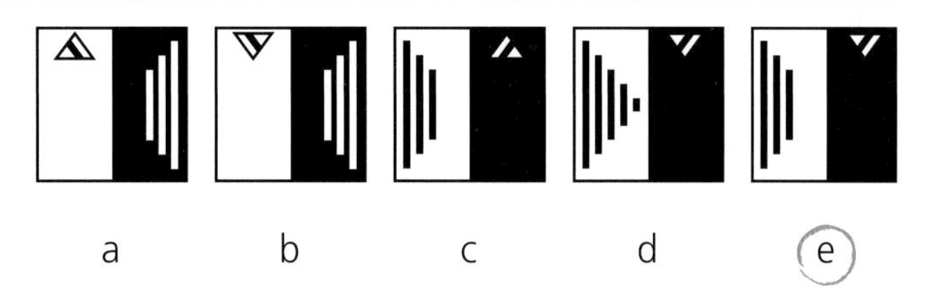

- These sequences are split, but they always connect in some way. Luckily, because the patterns are split out, they can be easier to spot and may not be as tricky as you think!
- Work **step-by step** through the features, first looking at the **top row**, then the **bottom row** and lastly **compare the rows**. In this example begin with the lines.
  - The number of lines **reduces** by one each time in the top row.
  - The number of lines **reduces** by one each time in the bottom row too.
  - The lines move left to right (looking between the top and bottom rows) with each pair having the same number of lines. The lines also change from black to white.

  These are two simple **repeating sequences** – imagine the bottom boxes sliding up between the top boxes if you find it easier to work out what is happening.
- Next, move on to the triangles.
  - They move **clockwise** around the rectangle and rotate **anticlockwise** in the top row.
  - They move **anticlockwise** around the rectangle and rotate **clockwise** in the bottom row.
  - They begin at the same point in the first pair and move the same distance in the opposite direction between boxes.
- As you are only being asked for a missing box on the top row, you only need to follow the top sequence to answer this question. So, there should be three black lines on the left and a triangle with thin diagonal lines at the top of the box, pointing downwards (remember the triangle is rotating anticlockwise). This makes the answer **e**.

Train

1 Here are the first four boxes of a sequence showing an amazing snake that can change colour and disappear! On a piece of paper, draw boxes and carry on the sequence to work out the answers to the questions below.

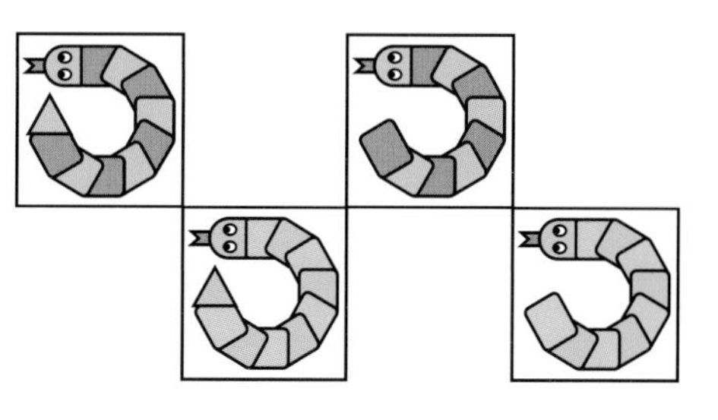

(a) In which box of the sequence will the last orange segment of the snake disappear? ________

(b) In which box of the sequence will both snakes disappear completely (the head, eyes and tongue all disappear at once)? ________

Try

The boxes on the left show a pattern that is arranged in a sequence. Choose the answer option that completes the sequence when inserted in the blank box. Circle the letter beneath the correct answer.

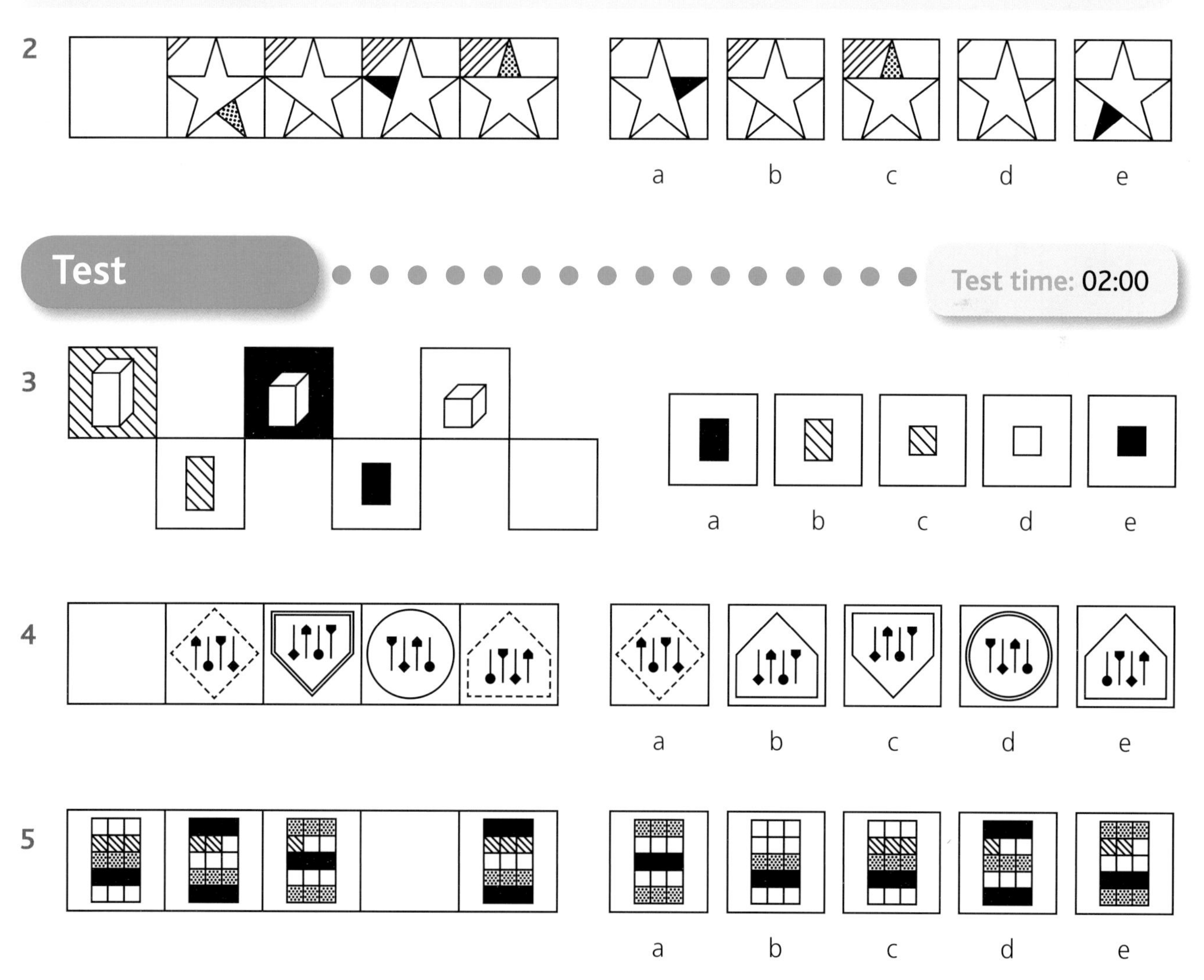

Test 5

Test time: 05:00

Complete this test in the time given above. Each question is worth one mark.

First you are shown a pattern that is arranged in a sequence. Choose the answer option that completes the sequence when inserted in the blank box. Circle the letter beneath the correct answer. Example:

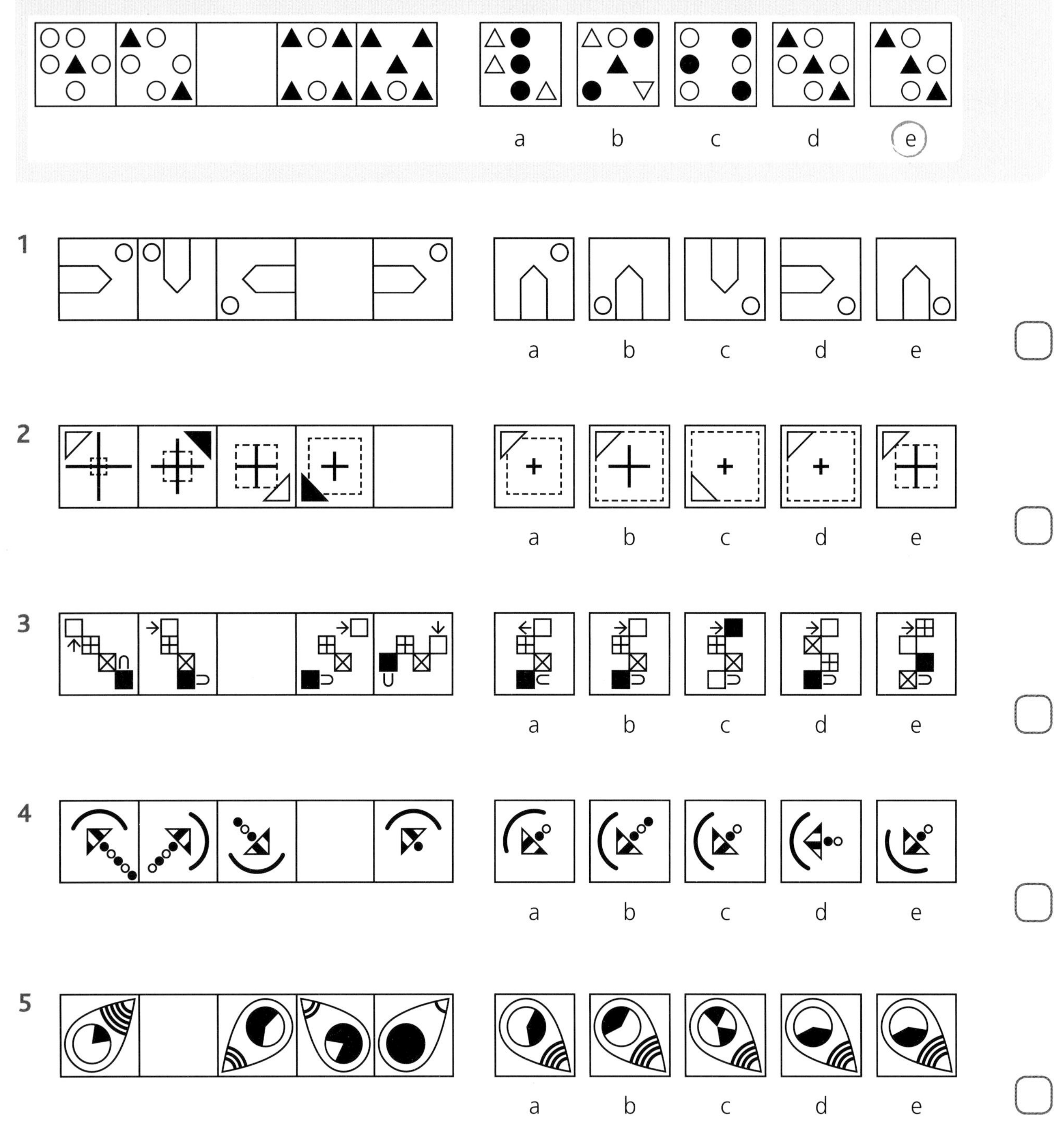

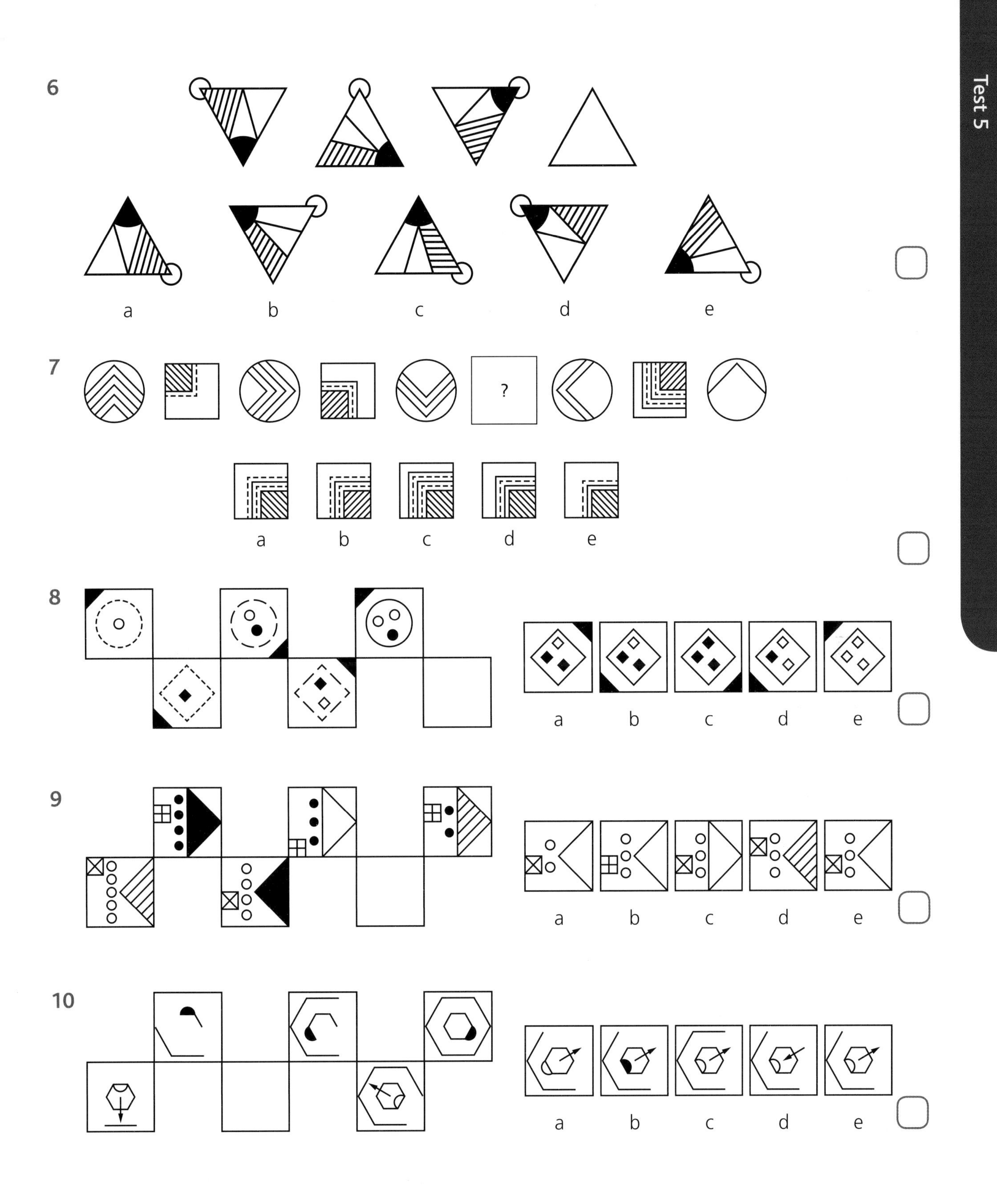

Record your score and time here and at the start of the book.

Score [ ] / 10 Time [ ] : [ ]

6 Matrices

Introduction

Now that you are used to working with patterns in a sequence, you are going to learn how to spot patterns that work in small grids called *matrices*.

You may have come across number squares in Maths and you will find that matrices work in exactly the same way. Patterns are created in rows and columns and can include any of the features you have worked with in this book.

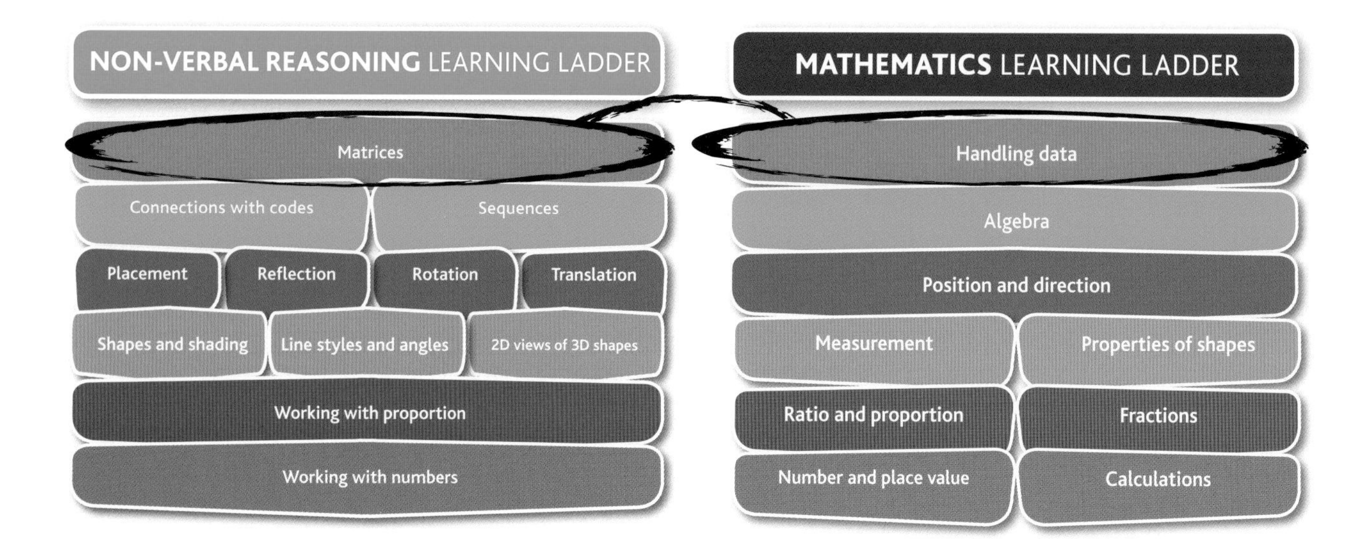

You have reached the top of the ladder so are now an expert in solving Non-Verbal Reasoning problems. Congratulations!

Just like in Maths, some of the trickiest questions in Non-Verbal Reasoning expect you to work with several pieces of data to spot hidden clues. The skill in solving these questions is in knowing where to look.

Matrices questions are all about looking at the **different directions** in the grids to spot where the patterns appear. Remember, the answer is always there to be discovered. You just have to know where to look.

Thinking skills and games

The following games can help to develop the skills featured in this chapter.

Flip it!

Create your own flipbooks using one of the free apps available. These electronic animations are an enjoyable way to practise moving images around for *Translation* and *Rotation* questions. They are fun to create as well!

Try using shapes to create images, such as triangles to make a tree, and find the best way to place the shapes in different layers to make the picture move about!

Advice for parents

Matrices questions involve a combination of the skills your child has already worked on throughout this book (although 3D questions are rare). These questions rely heavily on a logical ability to spot the direction of the patterns.

Most Non-Verbal Reasoning tests will include some variation of *Matrices* questions so they are worth taking time to master. As the patterns in *Matrices* questions can be generated in a number of different directions, they are a good way to assess children's ability to solve problems.

This chapter builds up an awareness of the various ways in which these patterns are created, as well as some of the common skills that are tested in these questions.

Any activities your children enjoy which involve creating patterns, such as decorating cakes or creating things using coloured strings and rubber bands, will help them to think about how patterns work. These activities can also be welcome distractions to reduce tension when they are feeling the pressure of approaching tests.

The games suggested in this chapter increase familiarity with transposing shapes (where a shape is moved from one position to another) as well as with reading information in rows and columns. Both skills are important in solving questions involving matrices.

Becoming familiar with where to look for clues comes with practice and there are many more examples of this question type in the *Workbooks* and *Practice Papers* accompanying this series.

Animal bingo

This is a game for two to four people. You will need coloured pencils and paper for each player. You will also need cards showing the following letters and numbers. Three of each are needed: 1A, 1B, 1C, 1D, 2A, 2B, 2C, 2D, 3A, 3B, 3C, 3D, 4A, 4B, 4C, 4D.

The aim of this game is to draw and spot a line of characters before anybody else!

1 Each player should draw the grid shown on their piece of paper, making it big enough so that they can draw a picture inside each box.
2 Shuffle the cards and place them face downwards.
3 Everybody takes a card and turns them over together. The players should all draw the image instructed by the card. For example, if the card they turned over is 3B, that player should draw a red chick in the square which is 2 down and 3 across. Put a time limit on the drawing of perhaps 30 seconds.
4 Continue to pick cards in the same way. The game finishes when somebody has a complete line or column, and shouts 'Bingo!' That player is out of the game if they have not drawn the correct animal or used the correct colour.
5 If all the cards are turned over without a winner, shuffle them again and place the pile face down. Carry on until a winner is found.

	1 Green	**2** Blue	**3** Red	**4** Yellow
A Cat				
B Chick				
C Bat				
D Sheep				

Matrices using numbers and shapes

Skill definition: Solve *Matrices* questions that use numbers and shapes.

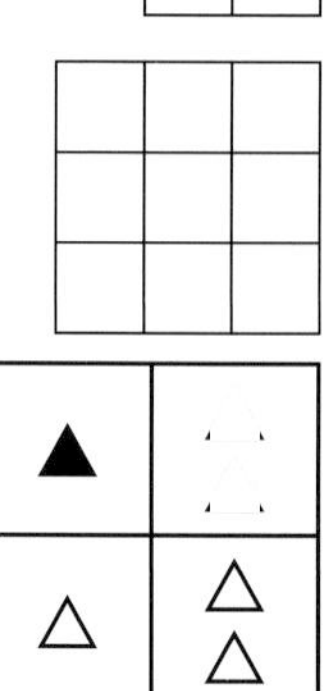

- Matrices are grids with patterns that work in different directions.
- In this learning spread you will practise questions with matrices that use **numbers**, **scale**, **proportion**, **2D and 3D shapes**, **shading**, **line styles** and **angles**.
- As explained on pages 22–23 in Chapter 1, most matrices have:
  - two rows and two columns (top picture), *or*
  - three rows and three columns (second picture).
- Patterns work **down** as well as **across,** and sometimes **diagonally**. They often include **short number sequences**. In the grid containing triangles shown here, the **rows** have an identical **shade** pattern and the **columns** have an identical **number** pattern.
- Matrices with three rows and columns often have **diagonal** patterns, as shown in this coloured grid. If a question has three different pictures on a single row, look for diagonal patterns. These should **repeat every third** diagonal. **Colouring** diagonal patterns can make it clearer to see how they work.

Method

One of the options on the right completes the pattern in the grid on the left. Circle the letter beneath the correct answer.

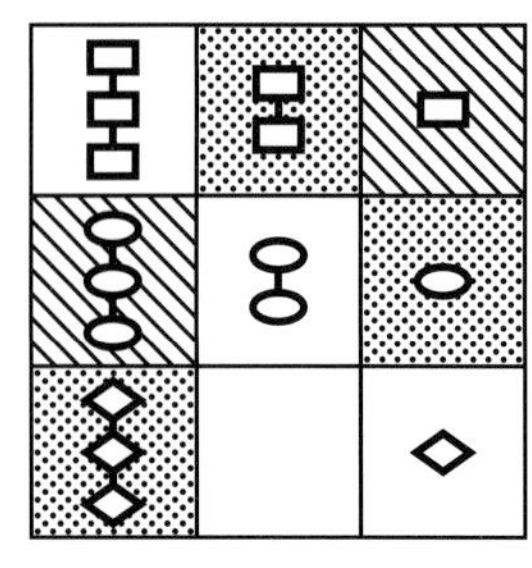

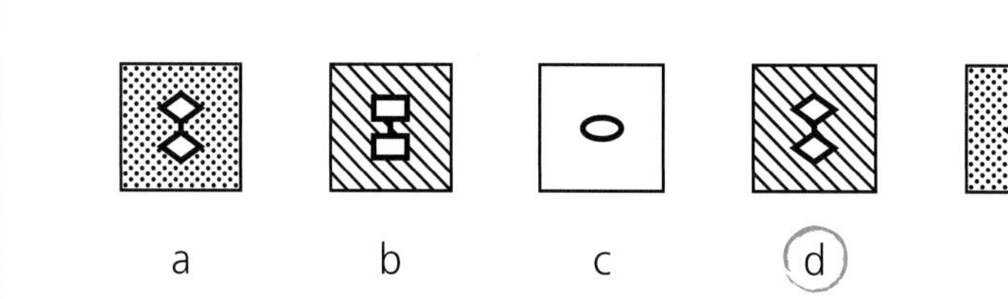

- Begin by breaking down the question into the different features to find the missing square. There seem to be patterns with **shape, number** of shapes and **shading**.
- Follow **one pattern at a time** to discover the rules:
  - the **shapes** are different on each **row**
  - the **number** of shapes changes in each **column** (decreasing by one from left to right)
  - the **shading** changes in a **diagonal** pattern working top left to bottom right.
- Now work out what **features** you would expect from the **missing square**:
  - the **shape** in the bottom row should be a **diamond**
  - the **number** of shapes in the middle column should be **two**
  - the **diagonal**, working top left to bottom right, that links to this empty box is **striped** (colour the diagonals, as described above if you find this difficult to see).

  This means you are looking for two diamonds with a diagonal background. So the answer must be option **d**.

Look out for distractors in questions. Not every feature in *Matrices* questions changes, like the oval in this question.

Train

1 There are two rules for this *Matrices* question: there is a different **fruit** in each **row** and the fruits change in **size** from one **column** to the next. Use your imagination to complete the missing squares with fruits of different sizes!

Try

One of the options on the right completes the pattern in the grid on the left. Circle the letter beneath the correct answer.

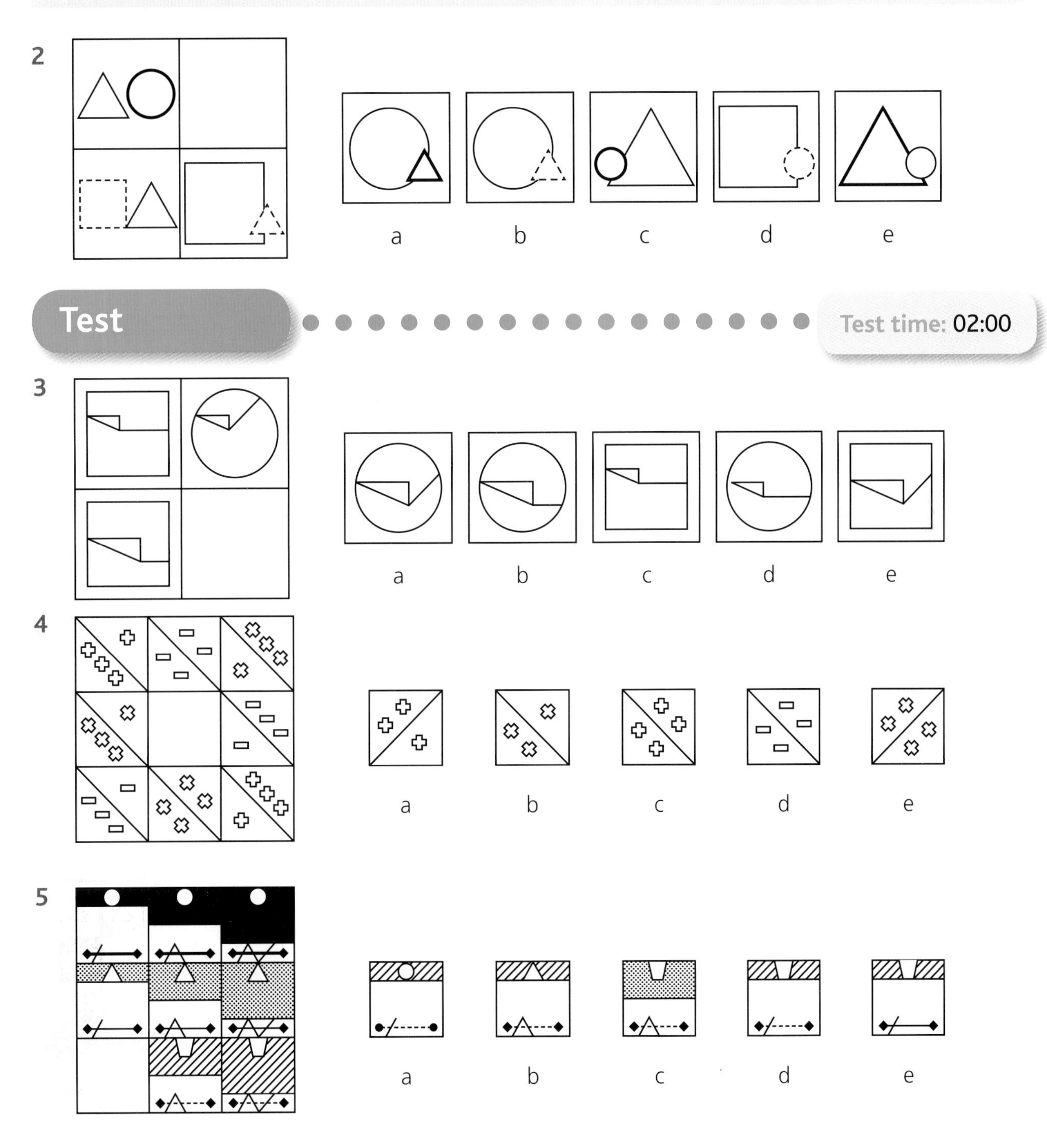

2

a b c d e

Test

Test time: 02:00

3

a b c d e

4

a b c d e

5

a b c d e

Matrices using position and direction

Skill definition: Solve *Matrices* questions that use position and direction.

- On this learning spread you will be solving more *Matrices* questions but, this time, using patterns with **layers**, **direction**, **reflection**, **rotation** and **translation**.
- You may see patterns:
  - **reflecting** in rows, columns and in the whole grid (as in the picture on the right)
  - rotating and changing direction in sequences.
- *Matrices* questions often use **translation** (see the example below), and spotting these patterns is the key to answering many of the trickiest questions.

Matrices questions can include patterns that run over more than one line. If you cannot see a pattern in the rows and columns, try looking at the next row for a connection. This cross moves half a side clockwise around the square. The pattern begins on row 1 going from left to right, and then continues on row 2 and then row 3.

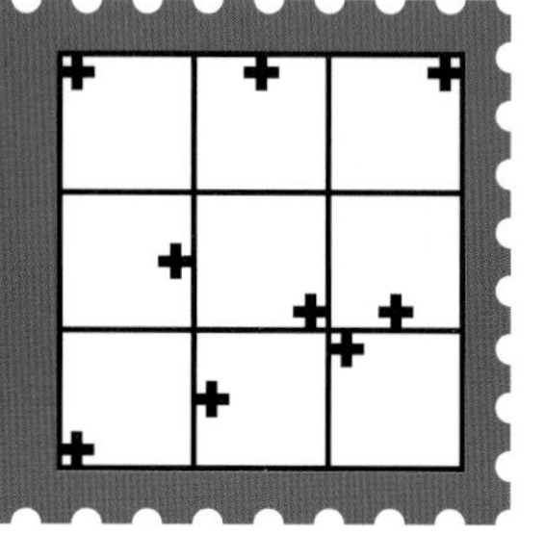

Method

One of the options on the right completes the pattern in the grid on the left. Circle the letter beneath the correct answer.

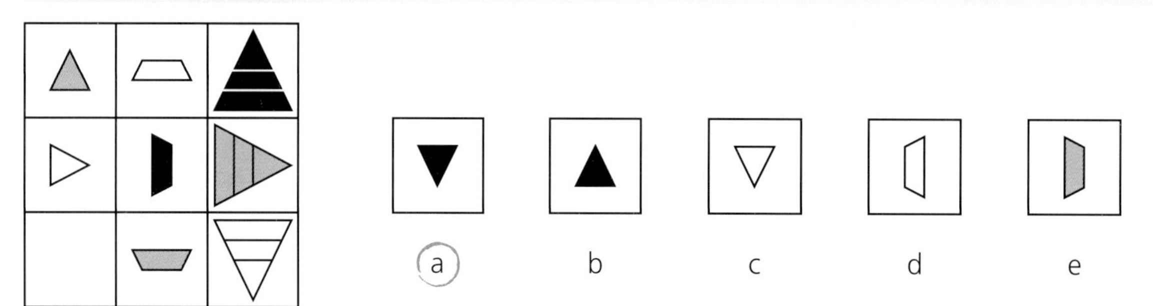

- Look for the patterns you can see in the question and try to work out what is going on with each one.
  - The shapes are different across each row, with the final shape made up of three pieces.
  - The shapes rotate 90° clockwise in the columns, working top to bottom.
  - The shading changes in a diagonal pattern working from top right to bottom left.
- When you see a different shape, as you do here, it is very likely to be a **translation**. Look carefully to see if the shapes in the first two columns are shown in the final column.
  - There is a **triangle** at the top and a **trapezium** beneath the **triangle**.
  - So you know that the right-hand shape is made up from the two left-hand shapes, plus an extra shape.
- Piece together what seems to be going on to work out the missing picture: the **shape** in the first column is a **triangle**; the **shading** of this diagonal is **black**; the final triangle is **pointing downwards**, so the answer must be option **a**.

Train

1 Work out the message in this puzzle by moving the letters around to make words.

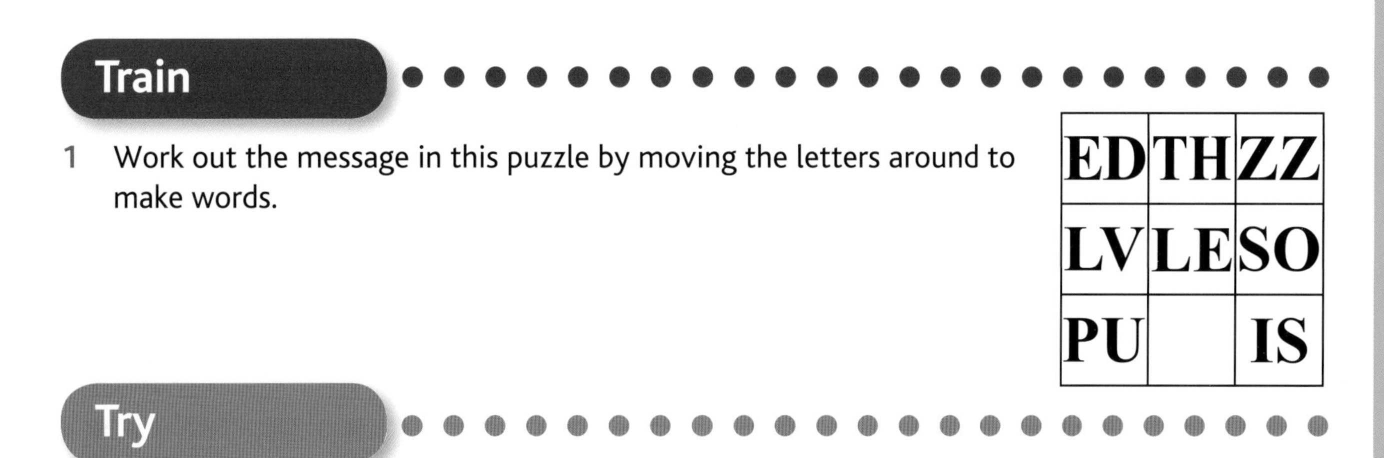

ED	TH	ZZ
LV	LE	SO
PU		IS

Try

One of the options on the right completes the pattern in the grid on the left. Circle the letter beneath the correct answer.

2

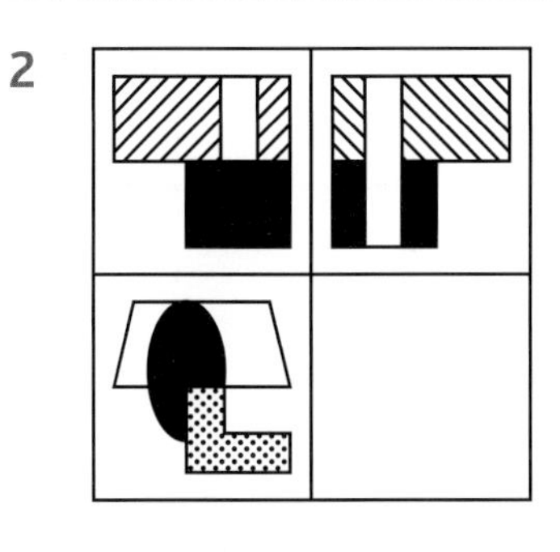

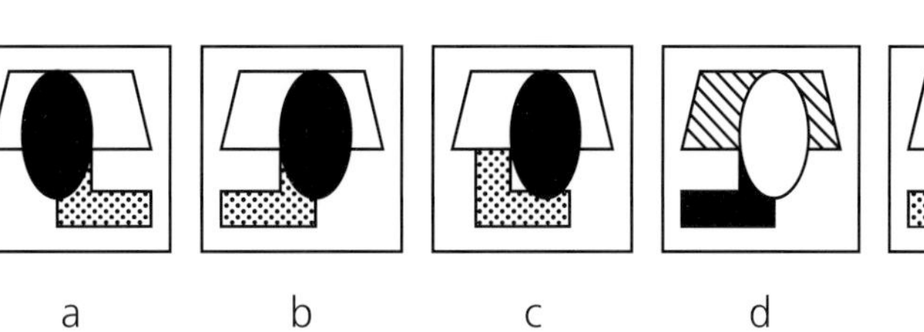

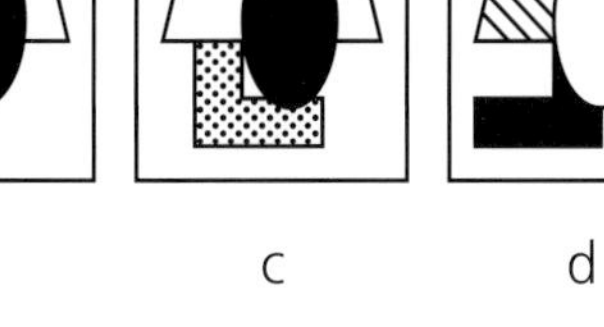

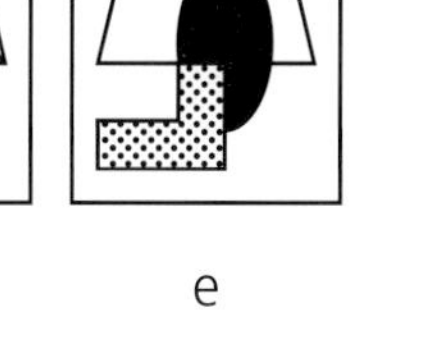

a b c d e

Test

Test time: 02:00

3

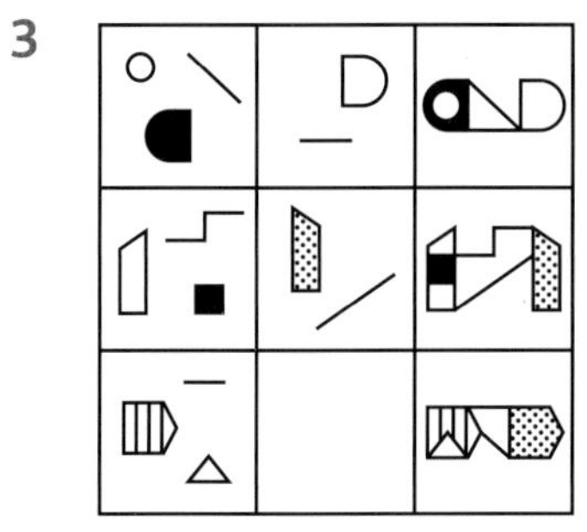

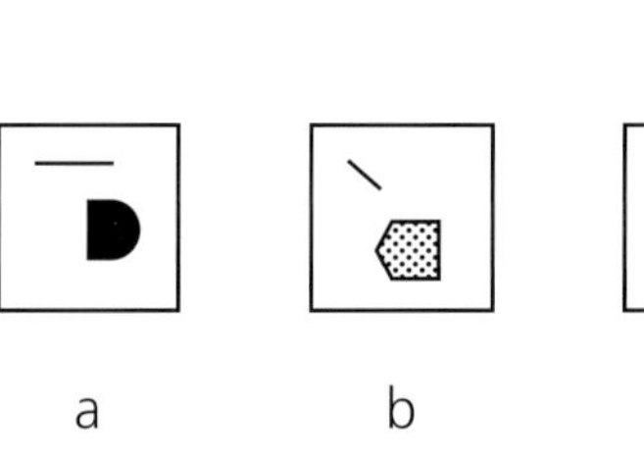

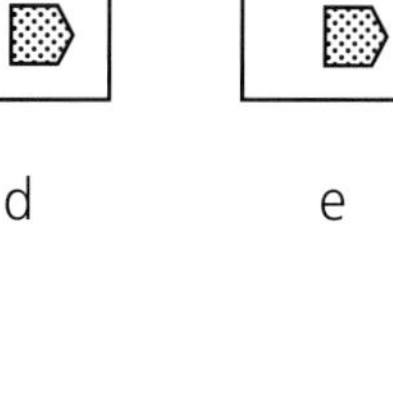

a b c d e

4

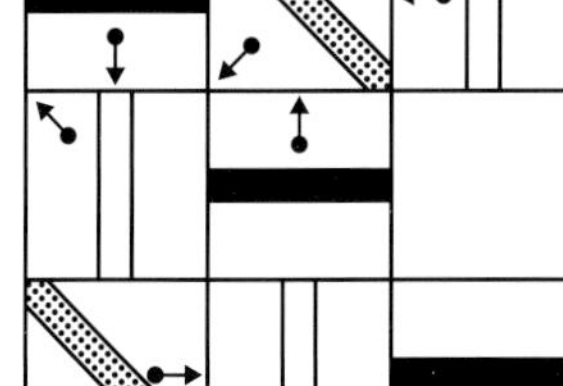

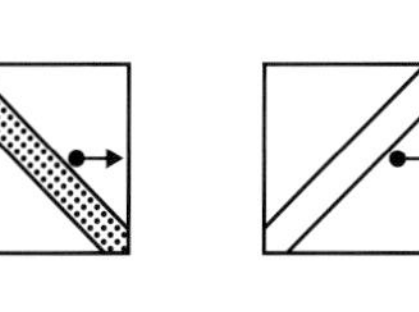

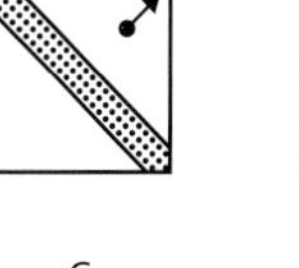

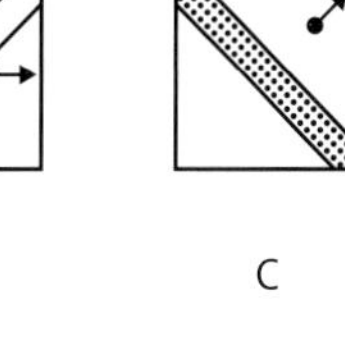

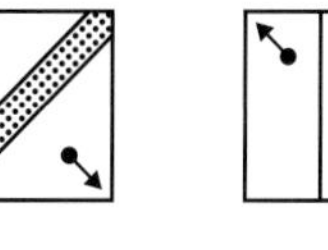

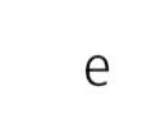

a b c d e

5

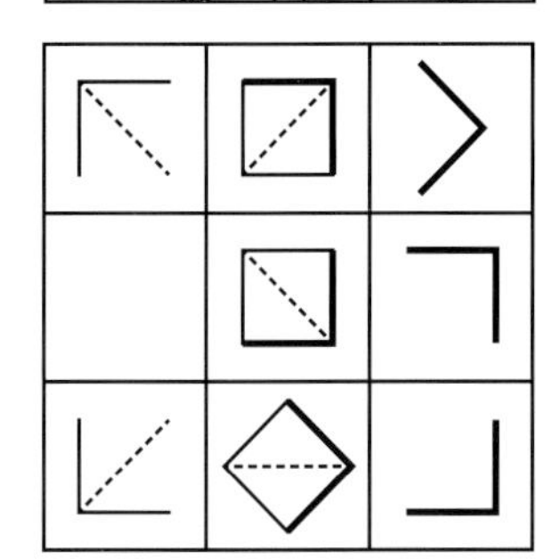

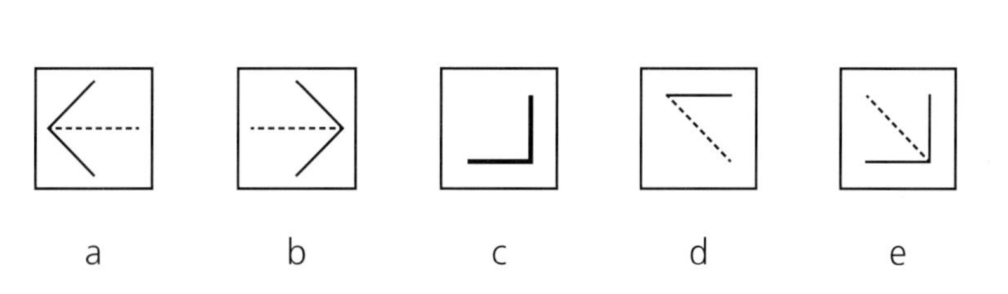

a b c d e

Matrices combining different skills

Skill definition: Solve *Matrices* questions arranged in different ways that use number, shape, position and direction.

- This is the last piece in your NVR jigsaw. You will combine all the skills you have learned in number, shape, position and direction to find out how to solve these final matrices questions.
- You may find *Matrices* questions that look a bit different.
  - Some may not have grid lines at all (as in the top picture on the right).
  - Others may use different shaped 'isometric' grids using triangles and hexagons (as shown in the second picture on the right).
  - You may even see grids with extra and missing squares, as in the example below (and described on page 23).
- The good news is that these variations are often simpler to work out. The symmetrical patterns can be easier to spot, and with missing boxes there are fewer things to look at!

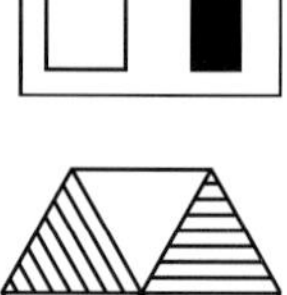

In *Matrices* questions that are difficult to solve, try exploring different directions. Some patterns work clockwise around a 2×2 grid while others might 'snake' around 3×3 grids.

Remember: Isometric grids often follow a circular pattern – see page 23.

Method

One of the options on the right completes the pattern in the grid on the left. Circle the letter beneath the correct answer.

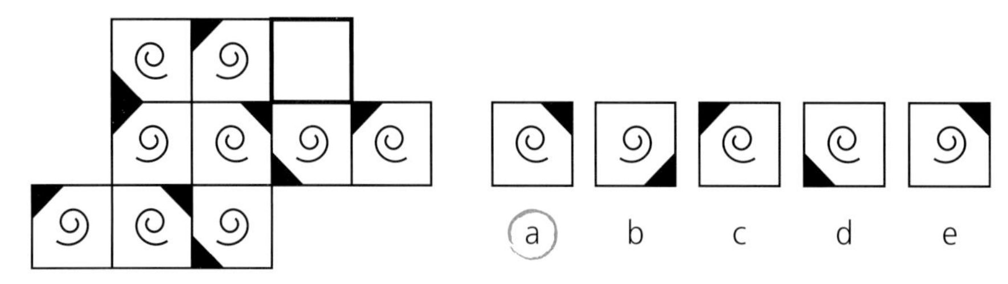

- The grid in this example looks a bit strange but it is clearly a matrix, just with one box removed at the bottom and extra boxes added to the sides. Do not be confused by the shape as the method is the same as that used in the other grids you have already seen already.
- Begin by looking for **patterns** that run **across**, **down** and in **diagonals**.
  - There is a simple sequence with the **curly shapes reflecting** in adjacent boxes across and down. This must mean that the missing shape has a curly shape a bit like a lower case letter 'e' (so the answer must be option **a**, **c** or **d**).
  - Looking from **top right** to **bottom left**, the triangle is in the same corner. This continues in the boxes outside the main grid too.
  - Colour the diagonals to make them clearer to follow.
  - This grid works in the same way as a standard 3×3 grid with every **third diagonal** being the same. The missing box is on the green diagonal so the answer must be option **a**.

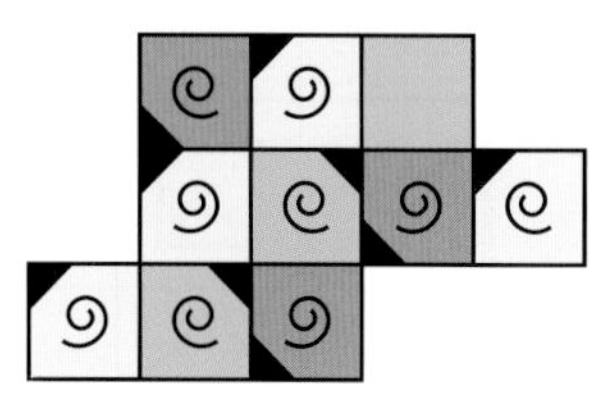

Train

1 Six of the triangles on the right are arranged to form a hexagon pattern with the shaded vertices in the centre. What is the black shape that will appear?

____

Try

One of the options on the right completes the pattern in the grid on the left. Circle the letter beneath the correct answer.

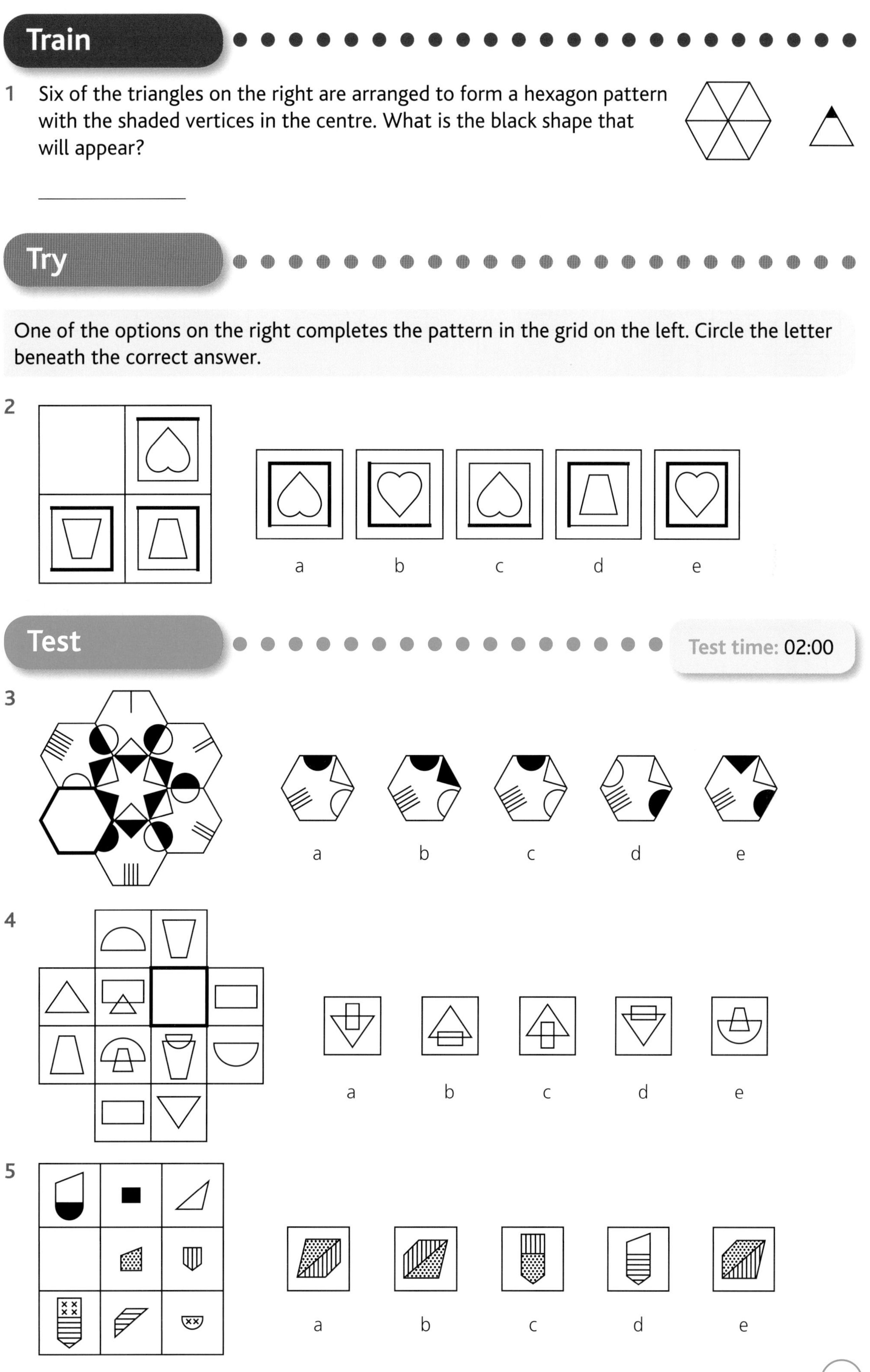

Test

Test time: 02:00

Test 6

Test time: 05:00

Complete this test in the time given above. Each question is worth one mark.

One of the options on the right completes the pattern in the grid on the left. Circle the letter beneath the correct answer. Example:

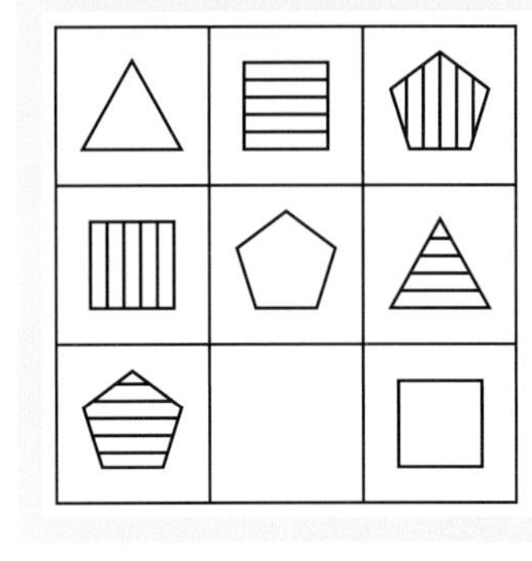

a b c d e

1

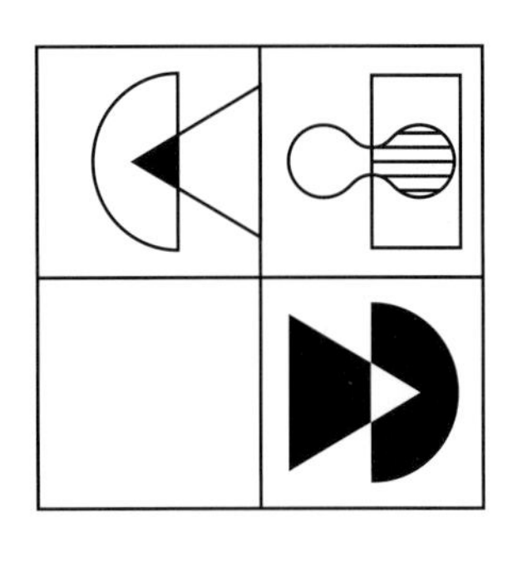

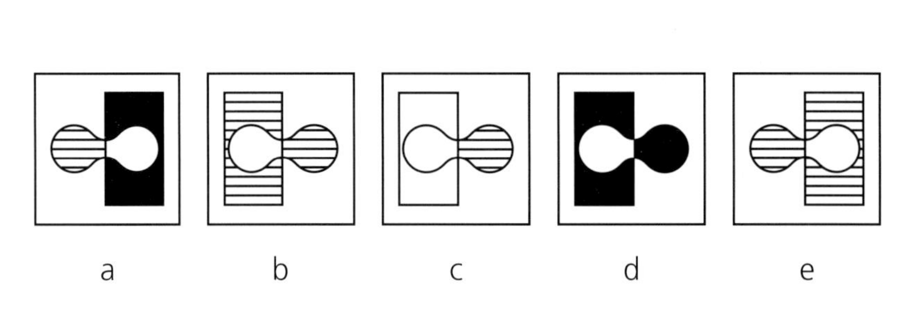

a b c d e

2

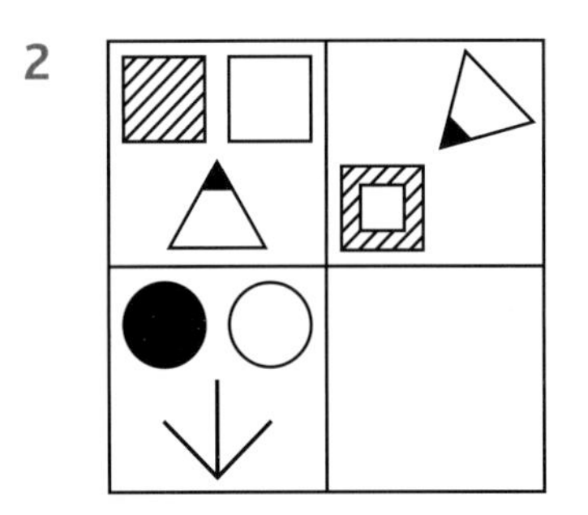

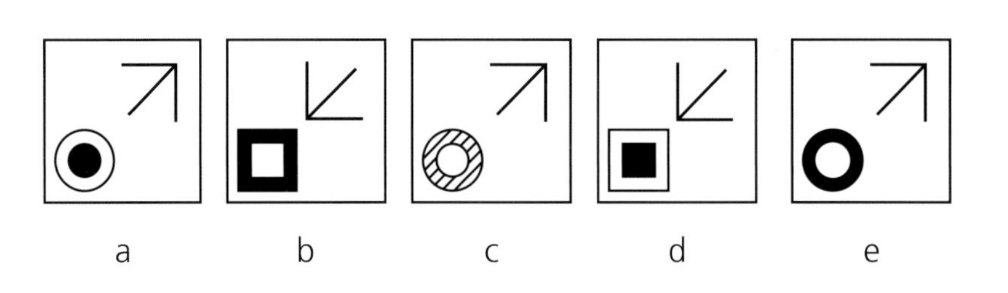

a b c d e

3

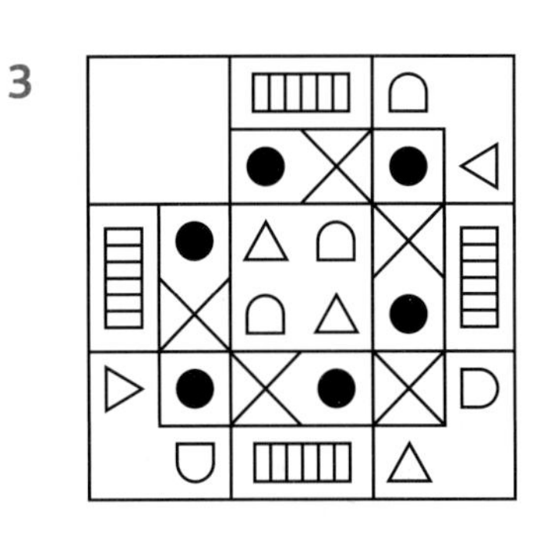

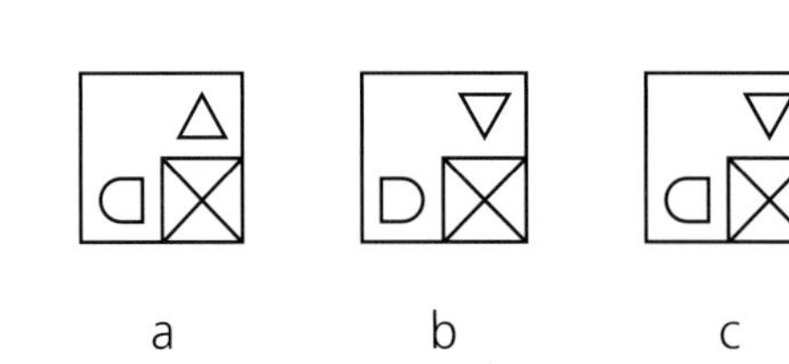
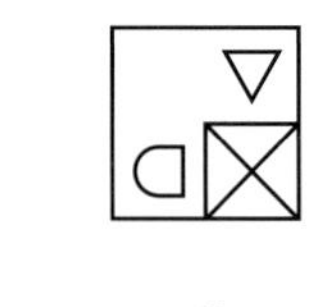
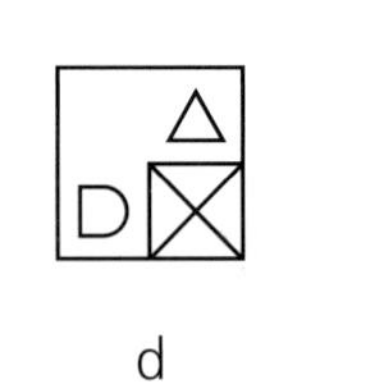
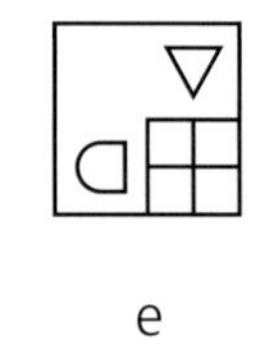

a b c d e

4

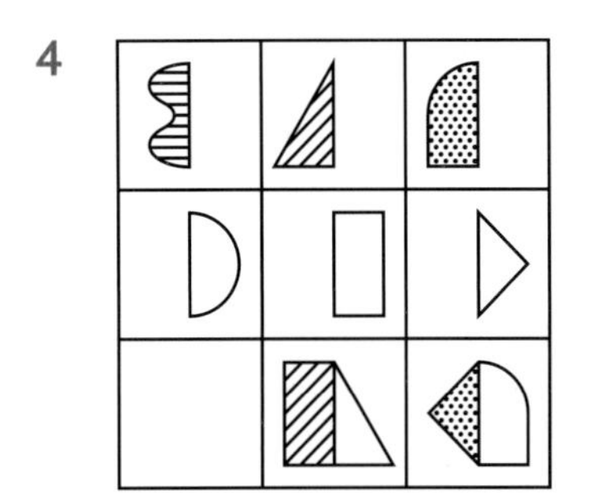

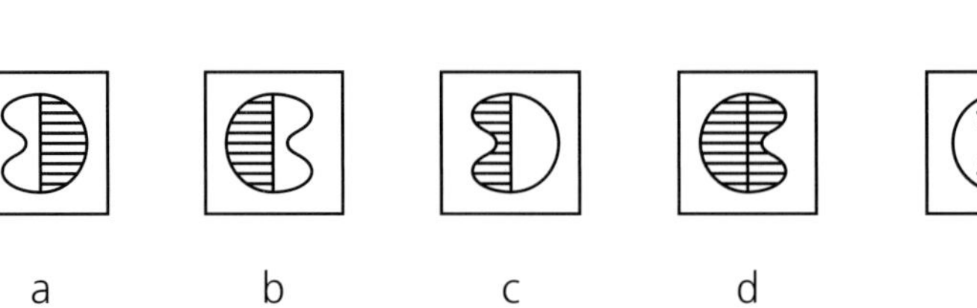
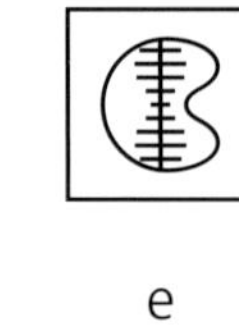

a b c d e

5

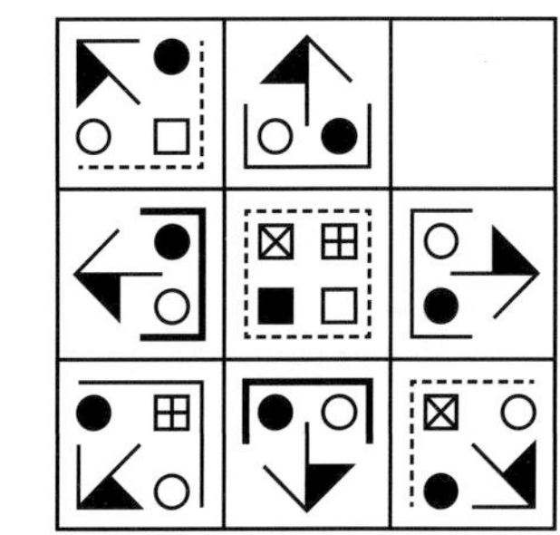

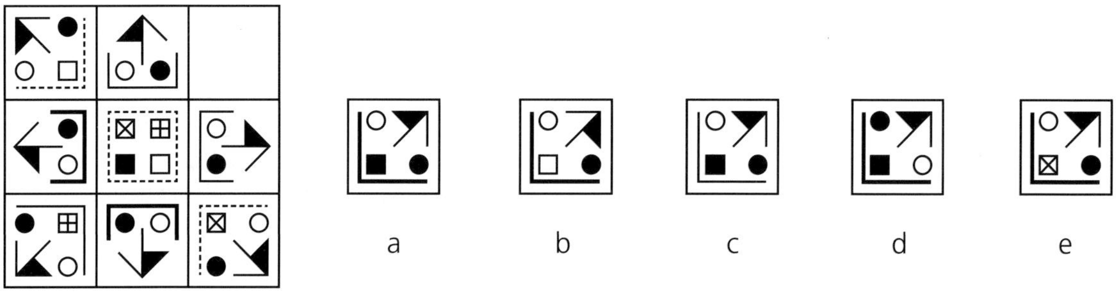

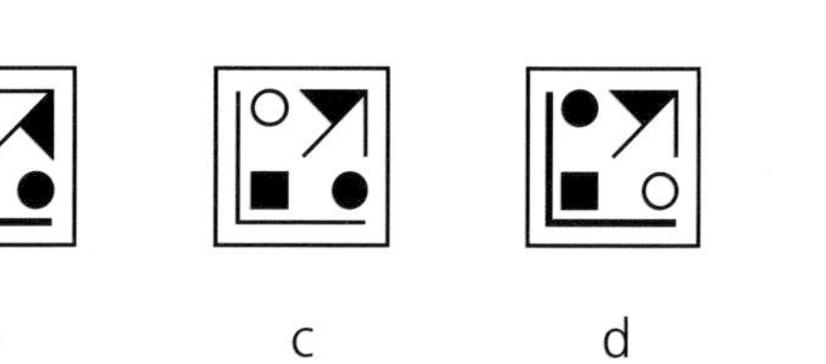

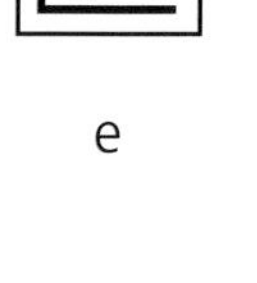

a b c d e

6

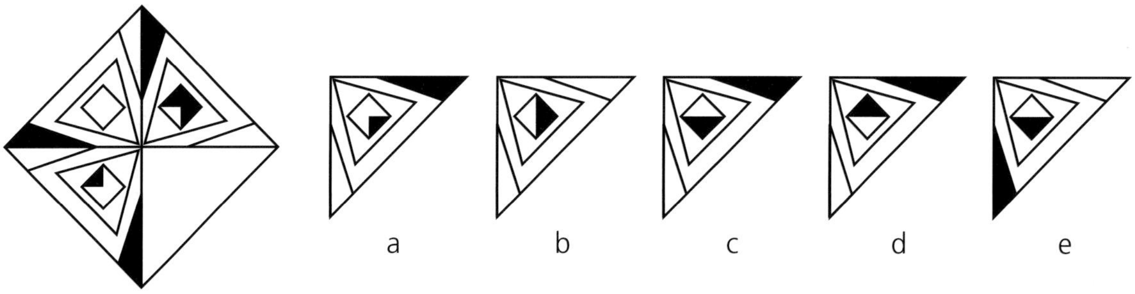

a b c d e

7

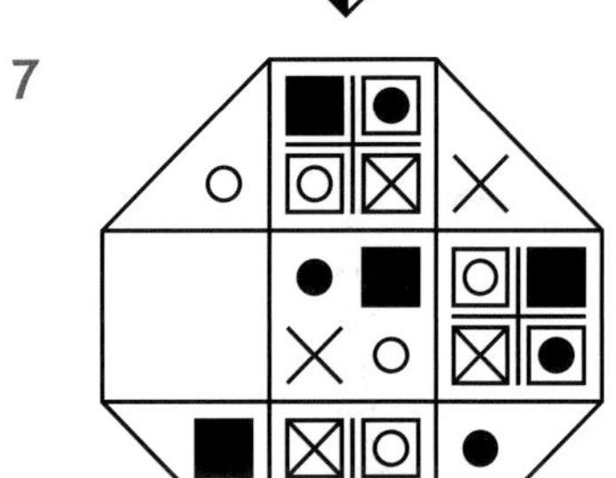

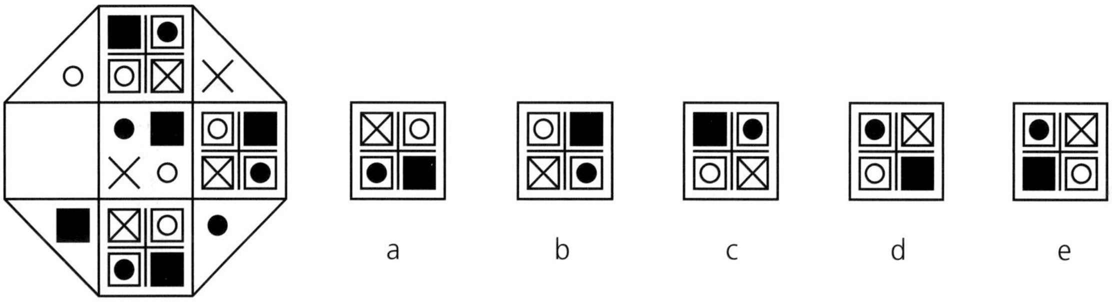

a b c d e

8

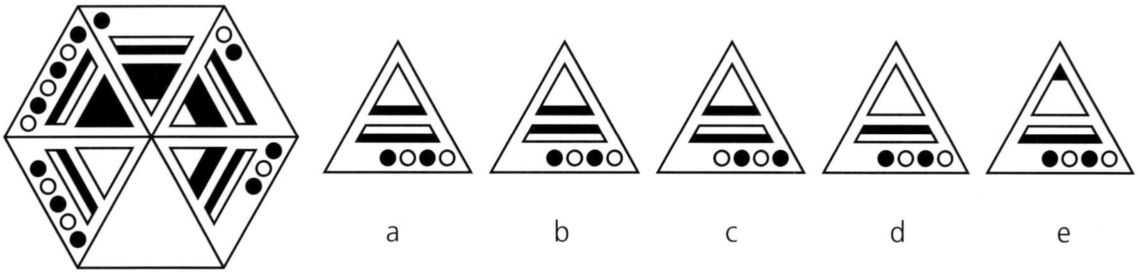

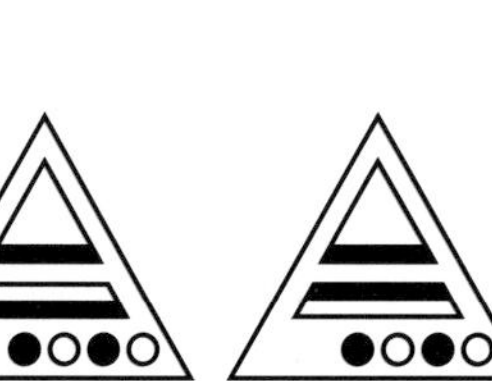

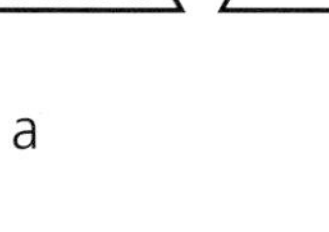

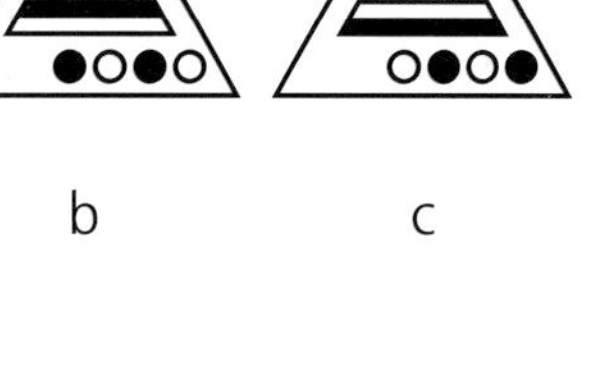

a b c d e

9

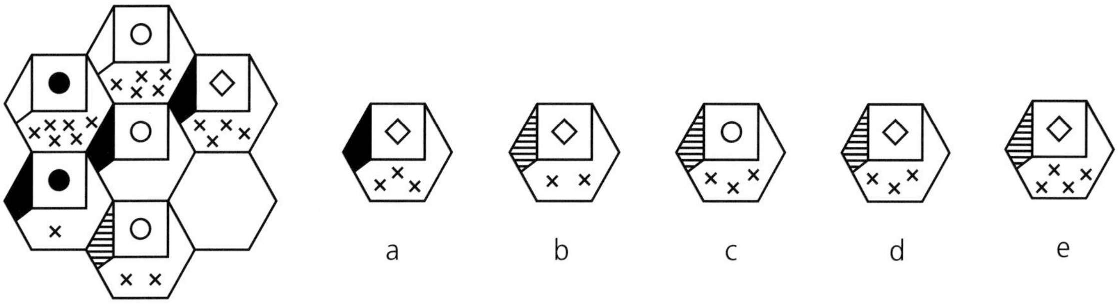

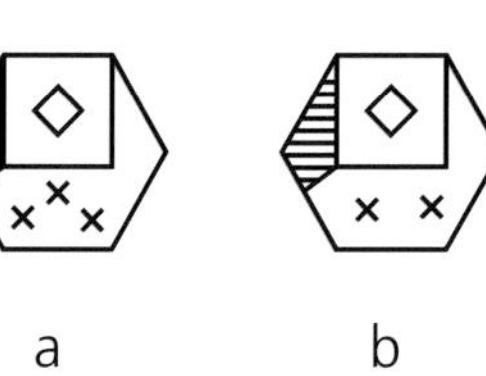

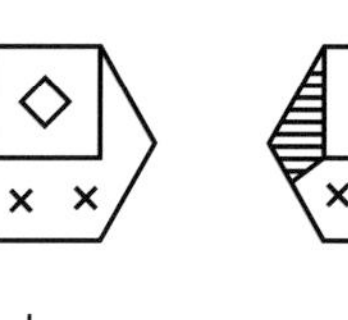

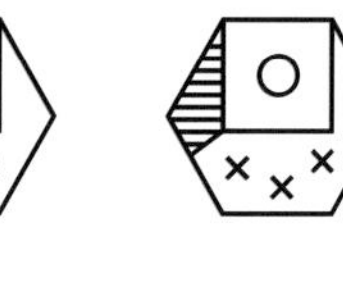

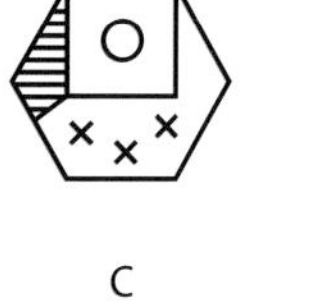

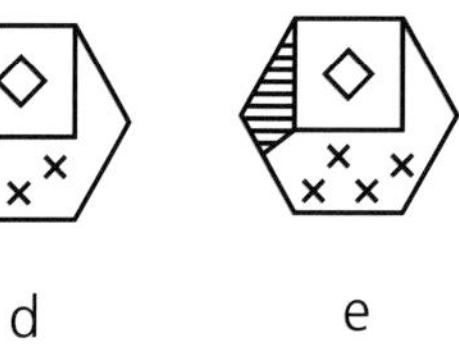

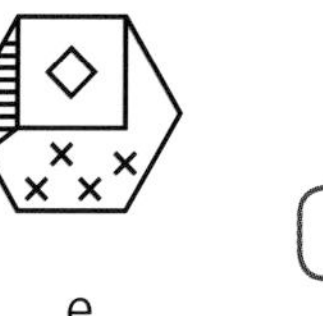

a b c d e

10

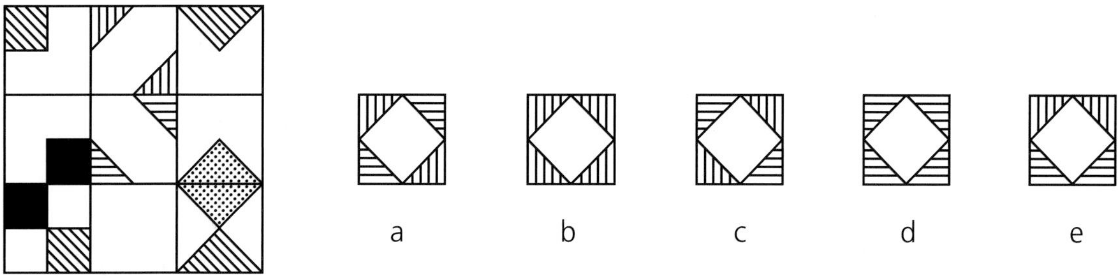

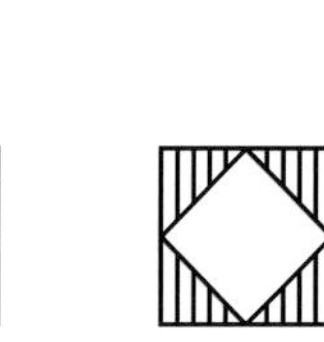

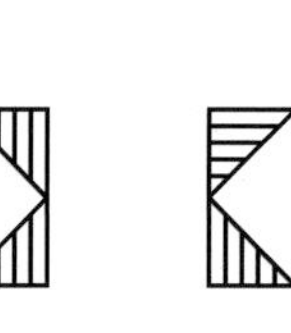

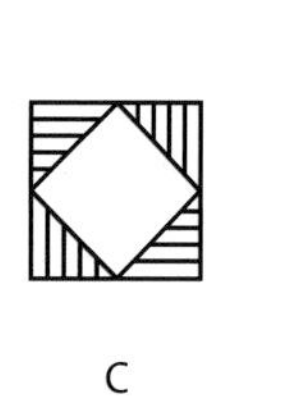

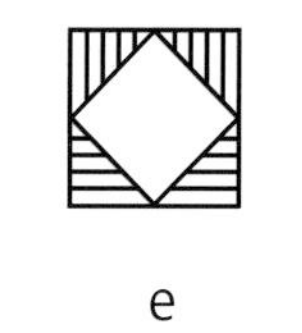

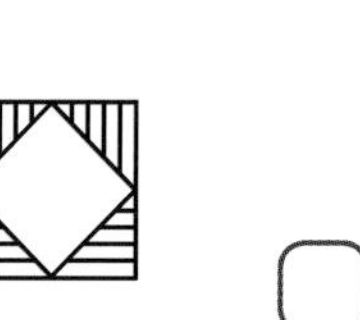

a b c d e

Record your score and time here and at the start of the book.

Score ☐ / 10 Time ☐ : ☐

11+ Sample test

Test time: 27:00

Complete this test in the time given above. Each question is worth one mark.

Look at these pictures. Identify the one that is most unlike the others. Circle the letter beneath the correct answer. Example:

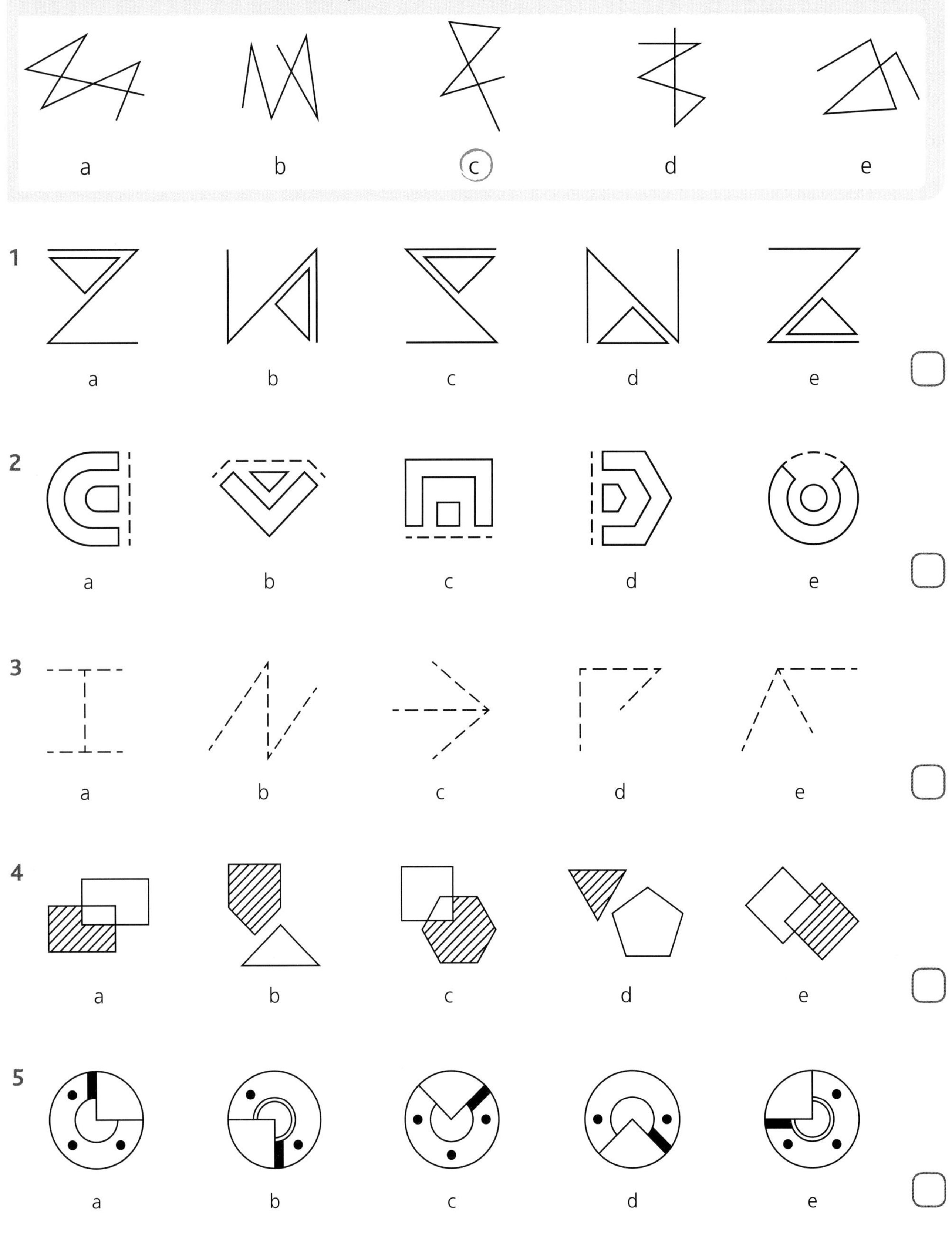

Look at the first two pictures and decide what they have in common. Then select the option from the five on the right that belongs in the same set. Circle the letter beneath the correct answer. Example:

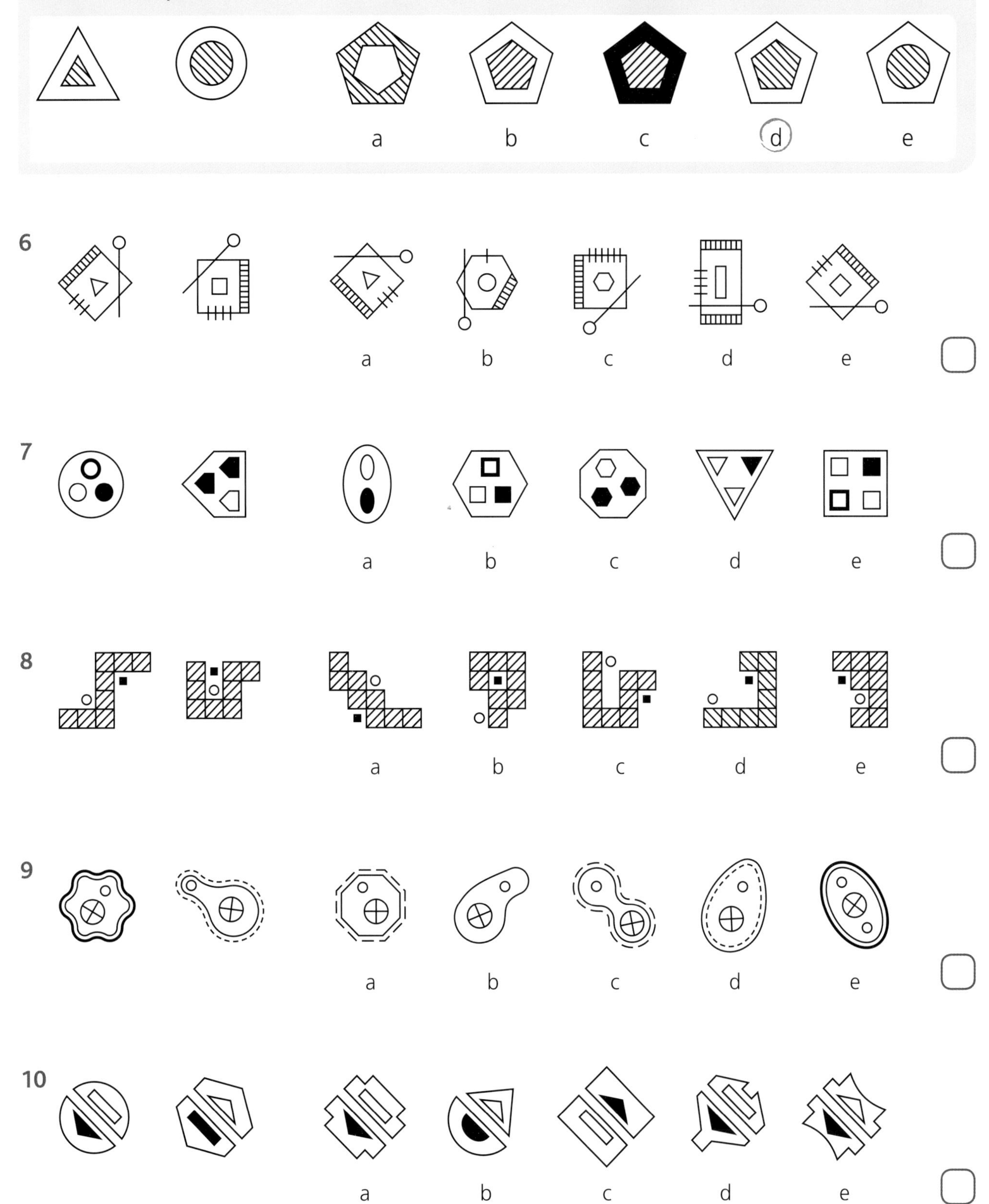

Look at the two pictures on the left connected by an arrow. Decide how the first picture has been changed to create the second. Now apply the same rule to the third picture and circle the letter beneath the correct answer. Example:

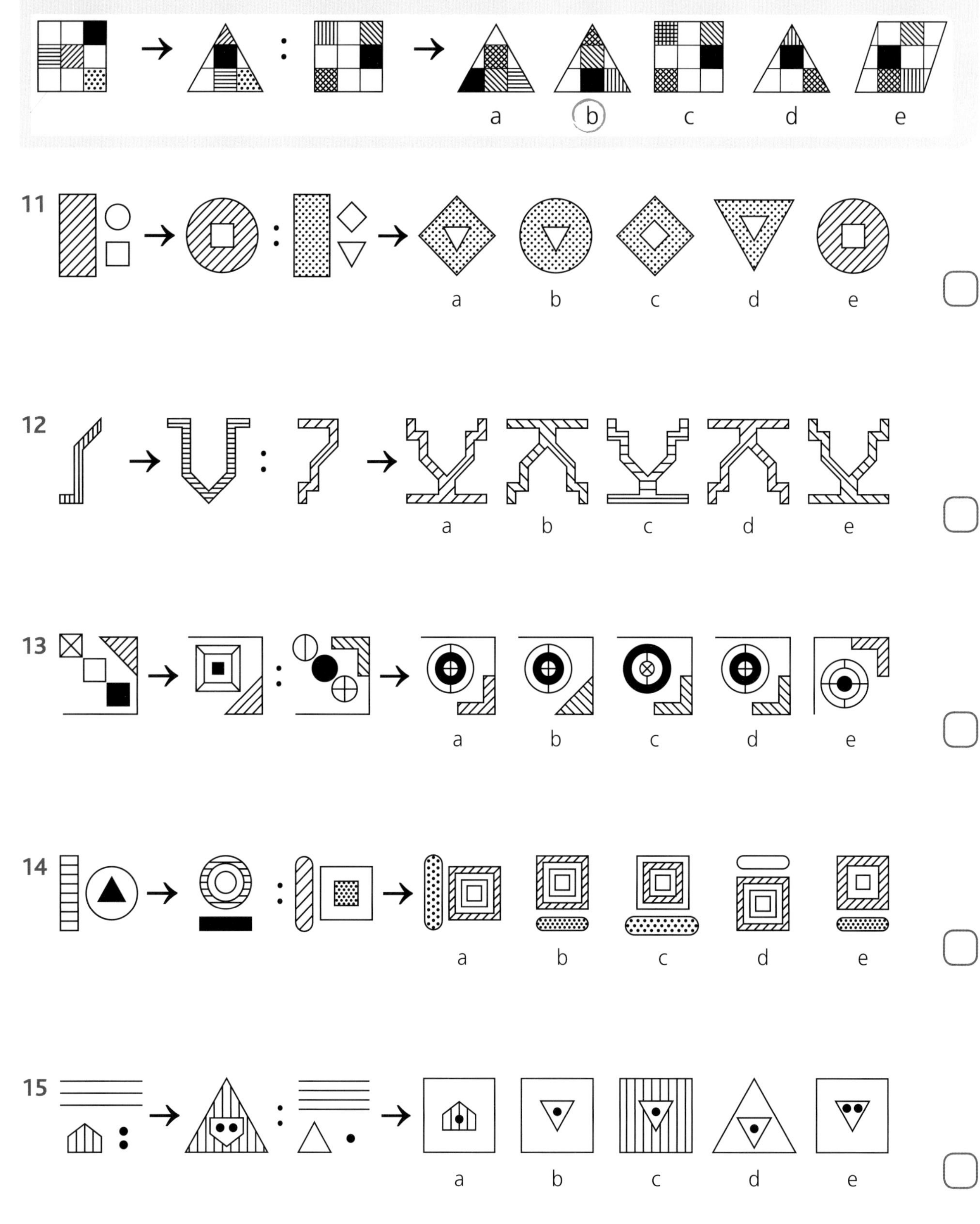

Find the cube, or other 3D shape, from the five options that can be made from the net shown on the left. Circle the letter beneath the correct answer. Example:

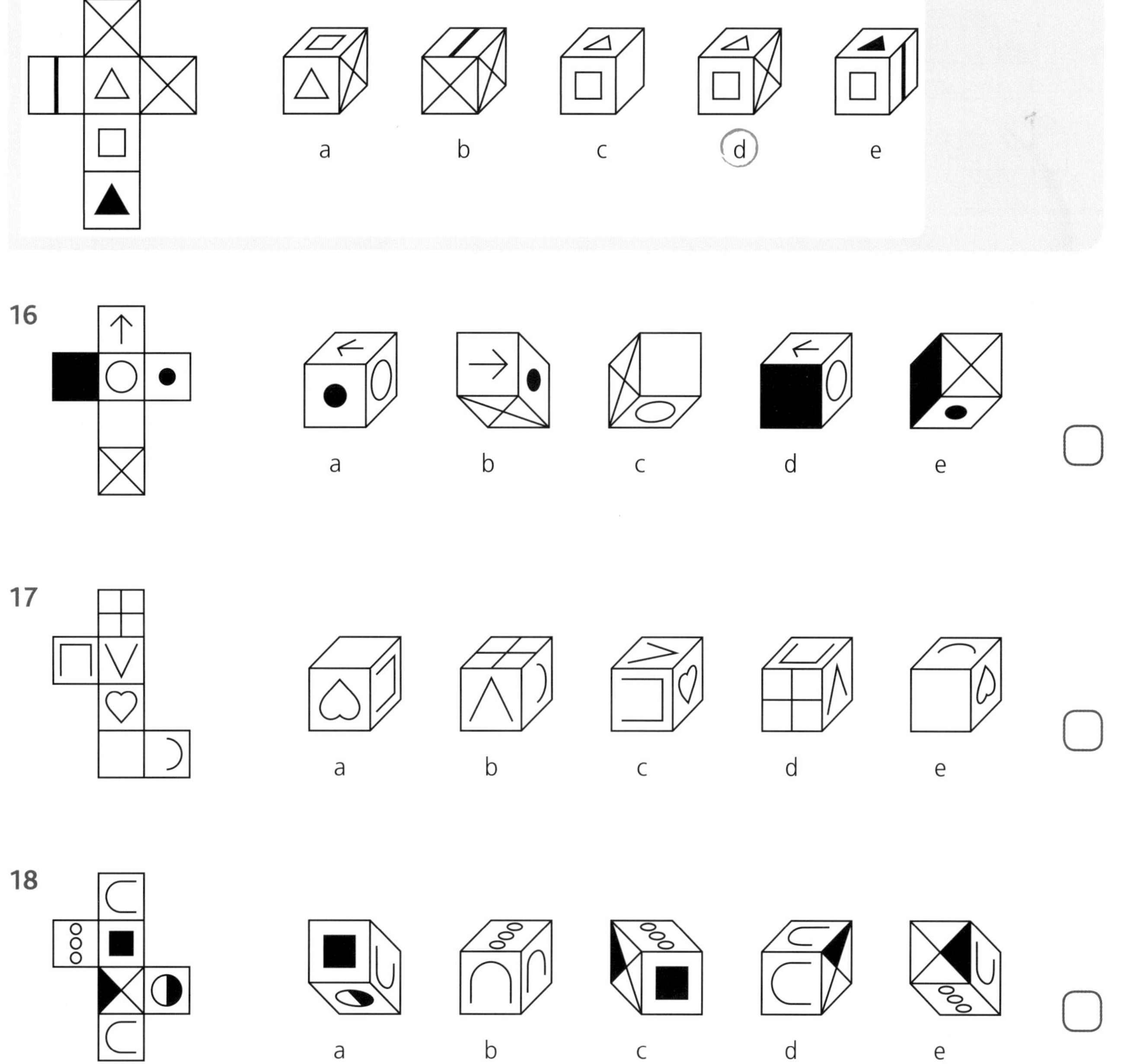

19 Choose the net that represents the shape on the left when it is unfolded. Circle the letter beneath the correct answer.

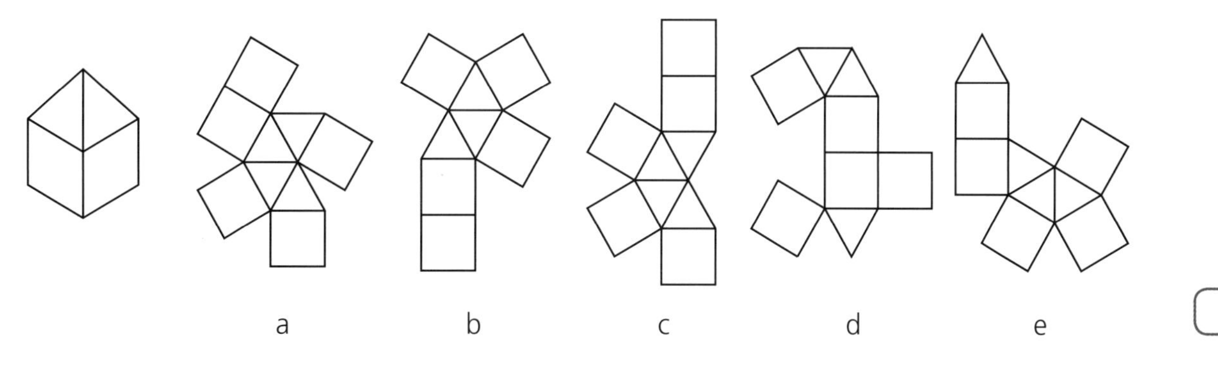

Which of the answer options is a 2D plan of the 3D picture on the left, when viewed from above? Circle the letter beneath the correct 2D plan. Example:

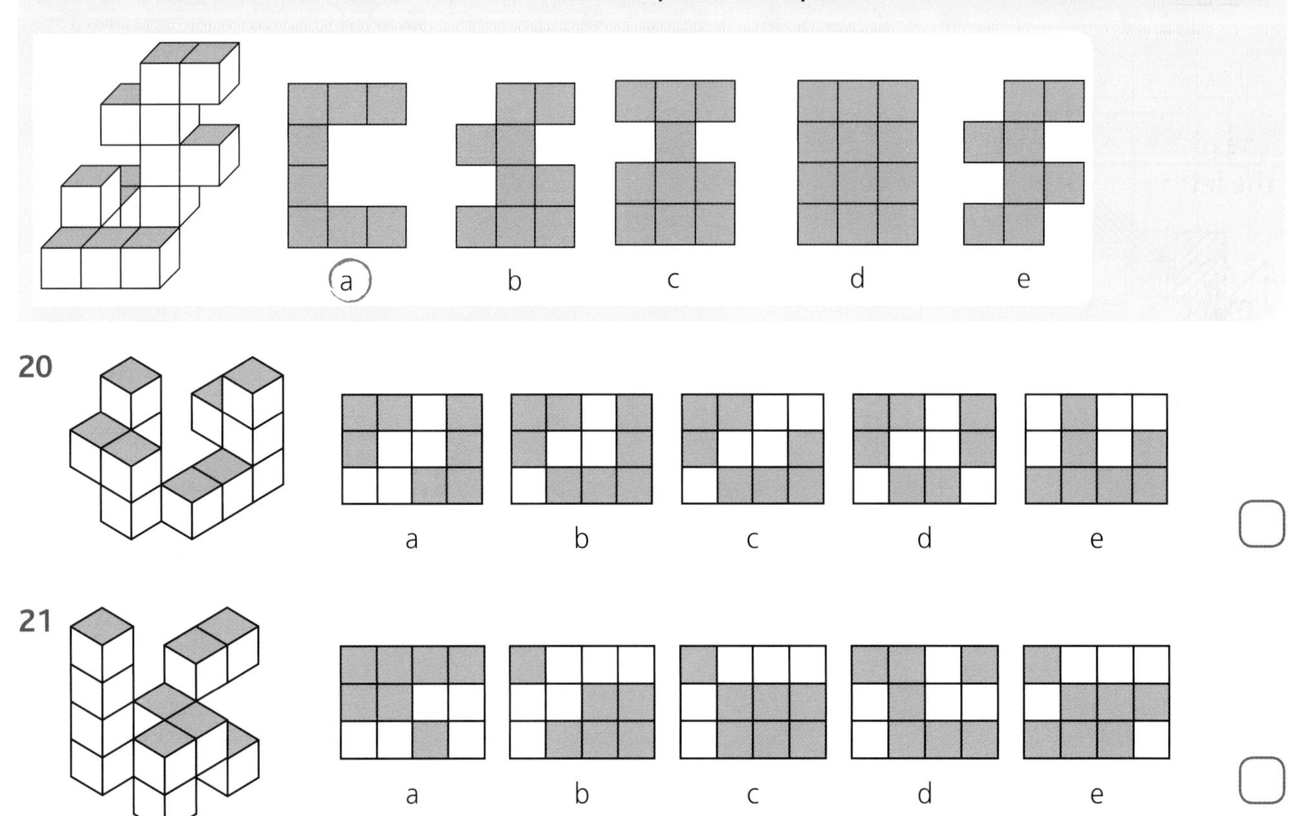

The square given at the beginning is folded in the way indicated by the arrows, and then holes are punched where shown on the final diagram. Identify the answer option which shows what the square would look like when it is unfolded. Circle the letter beneath the correct answer. Example:

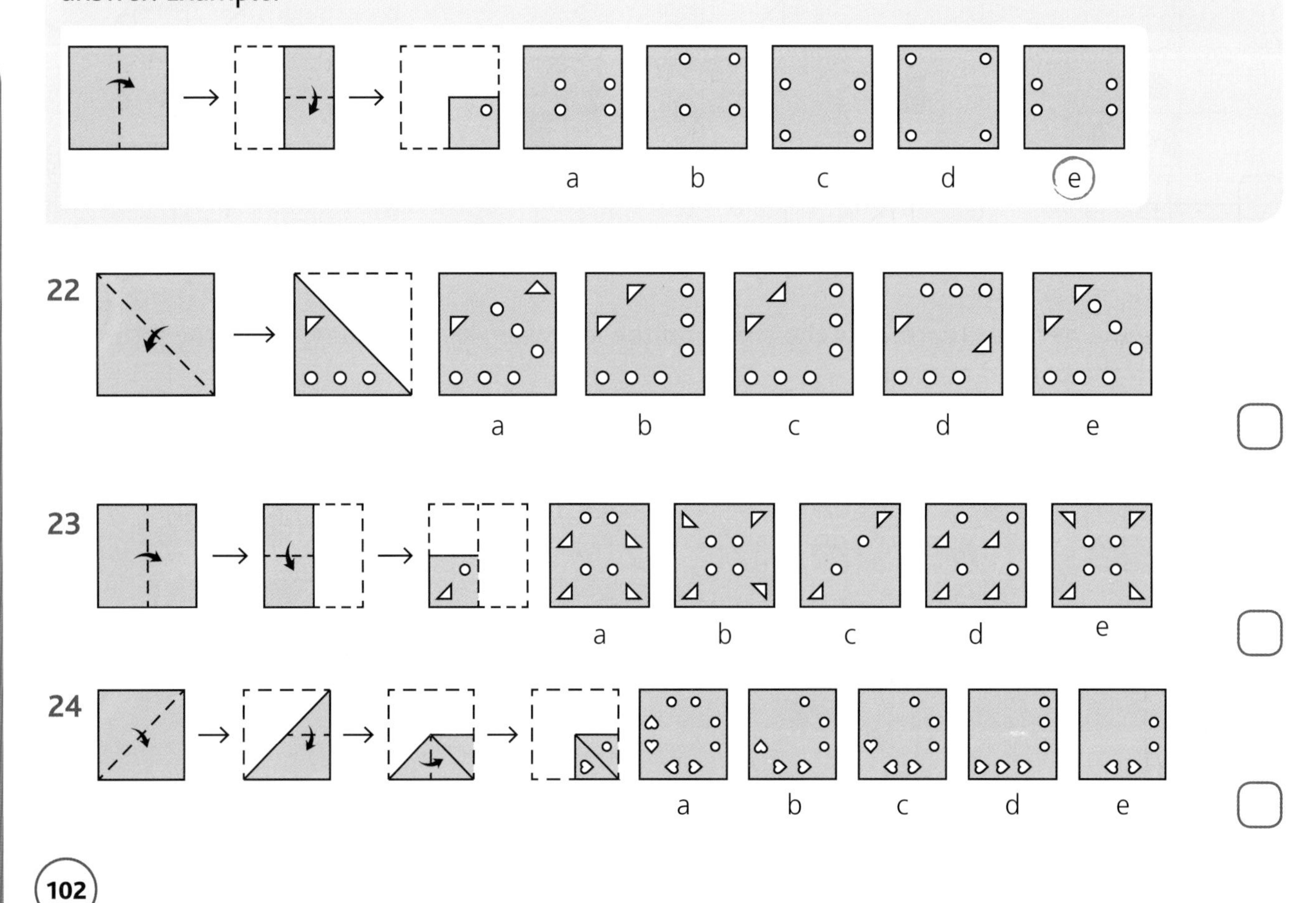

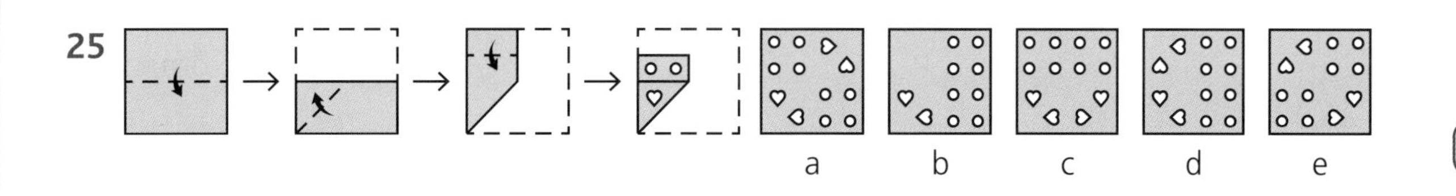

The picture on the left is rotated and is represented by one of the pictures on the right. Circle the letter beneath the correct answer. Example:

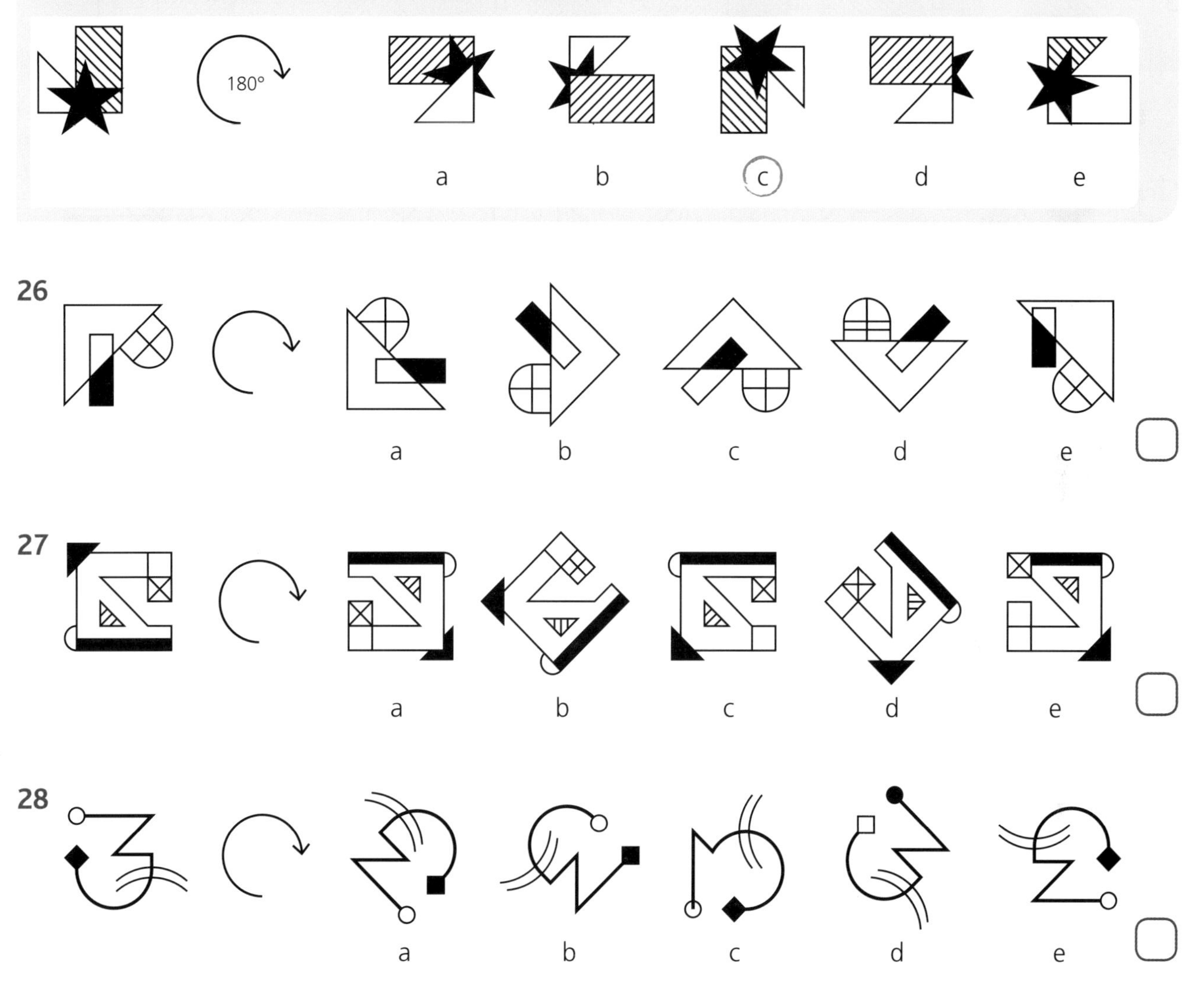

The picture on the left is reflected in a vertical mirror line and is represented by one of the pictures on the right. Circle the letter beneath the correct answer. Example:

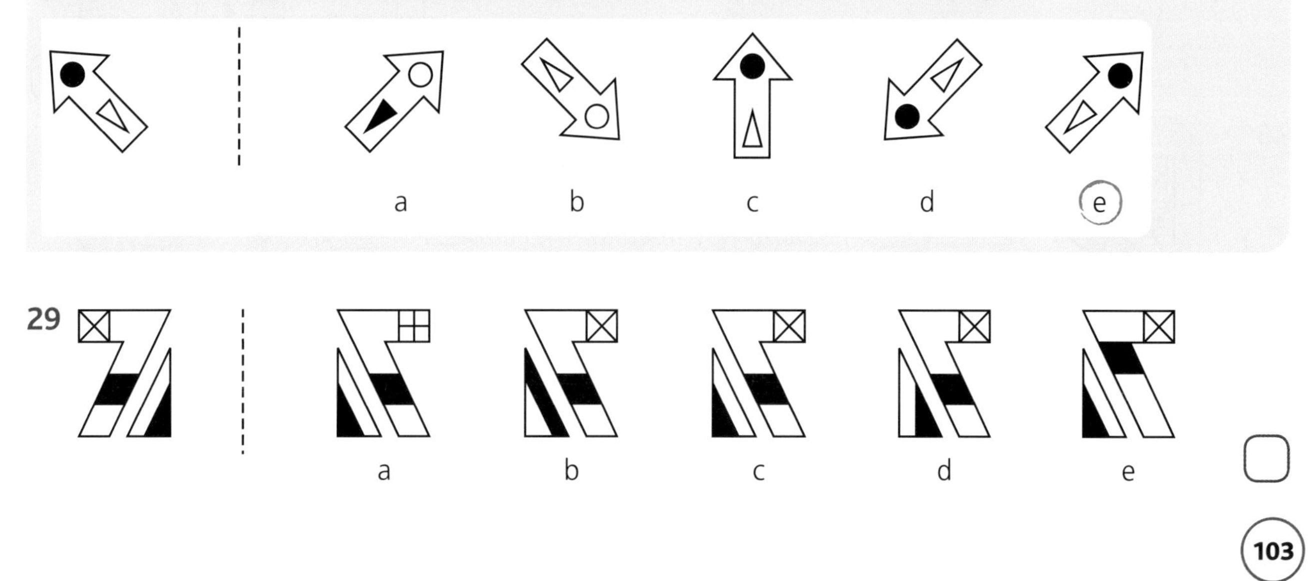

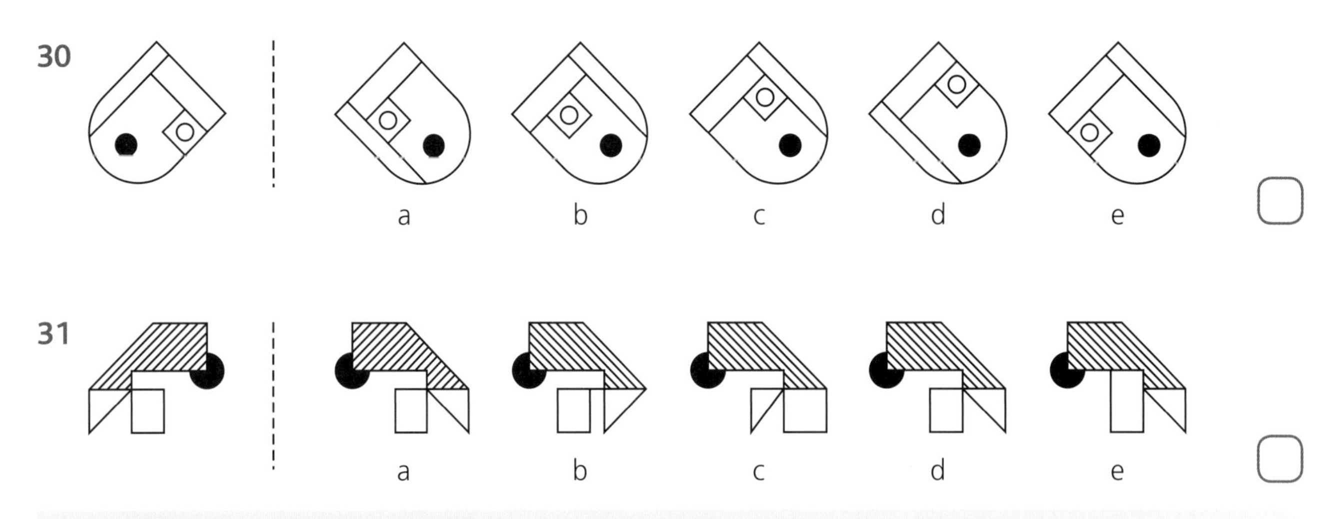

30 a b c d e

31 a b c d e

In questions 32 to 36 you will see a rotated version of one of the 3D diagrams (**a**, **b**, **c**, **d**, **e**). Circle the letter of the answer option that indicates the matching diagram above.

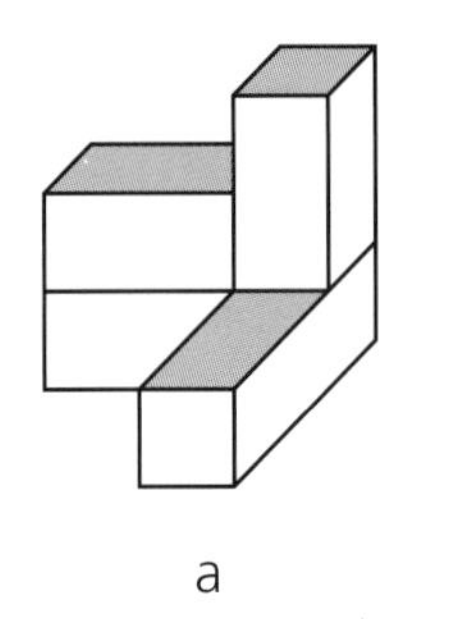
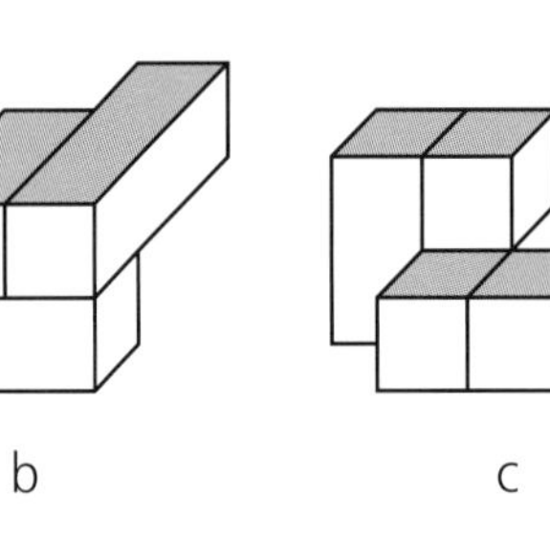
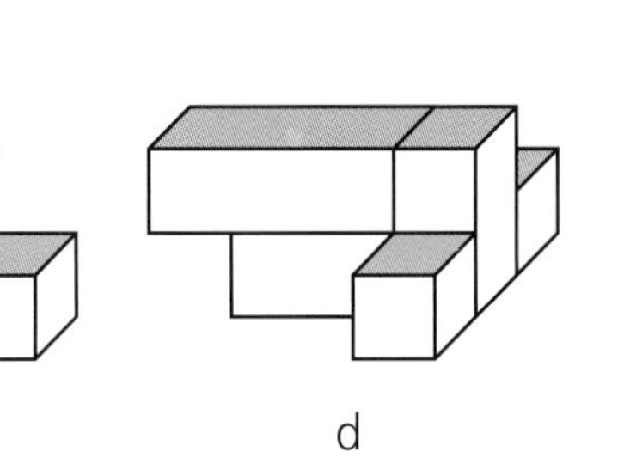
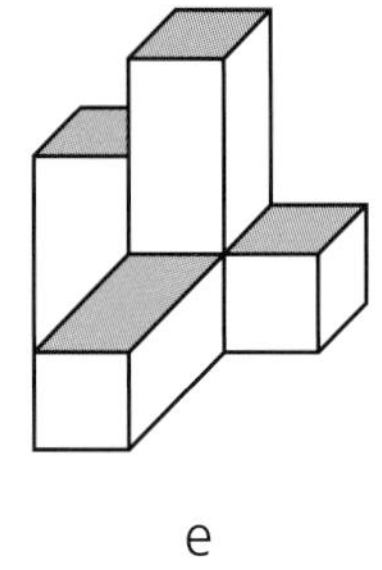

a b c d e

32

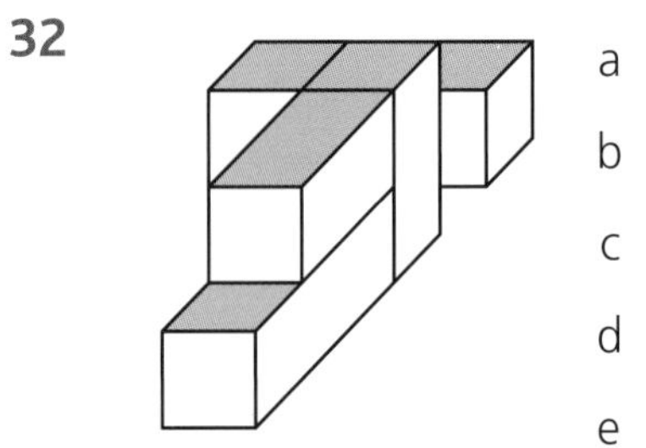

a
b
c
d
e

33

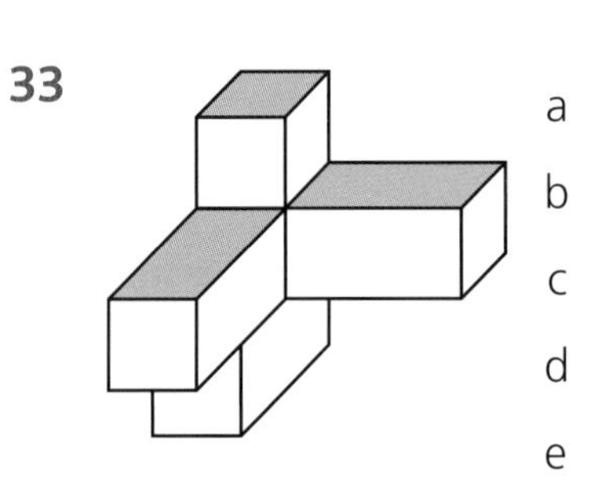

a
b
c
d
e

34

a
b
c
d
e

35

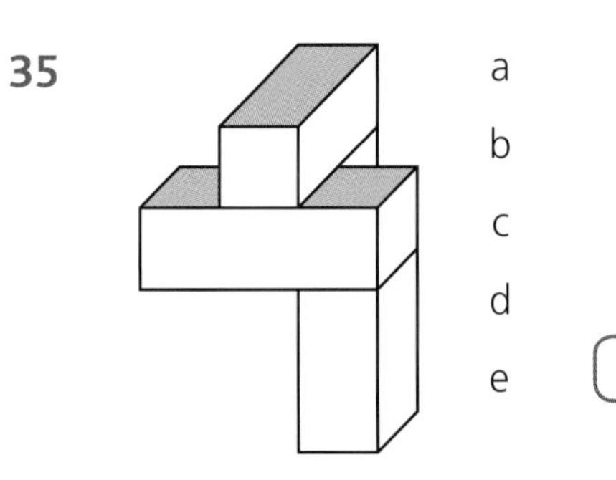

a
b
c
d
e

36

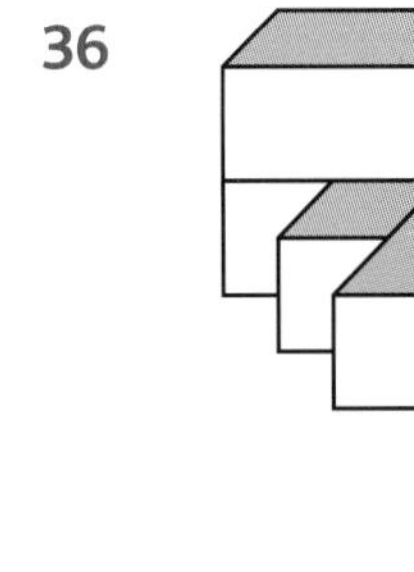

a
b
c
d
e

The picture on the left has been rotated clockwise by the number of degrees shown to give one of the pictures on the right. Circle the letter beneath the correct answer. Example:

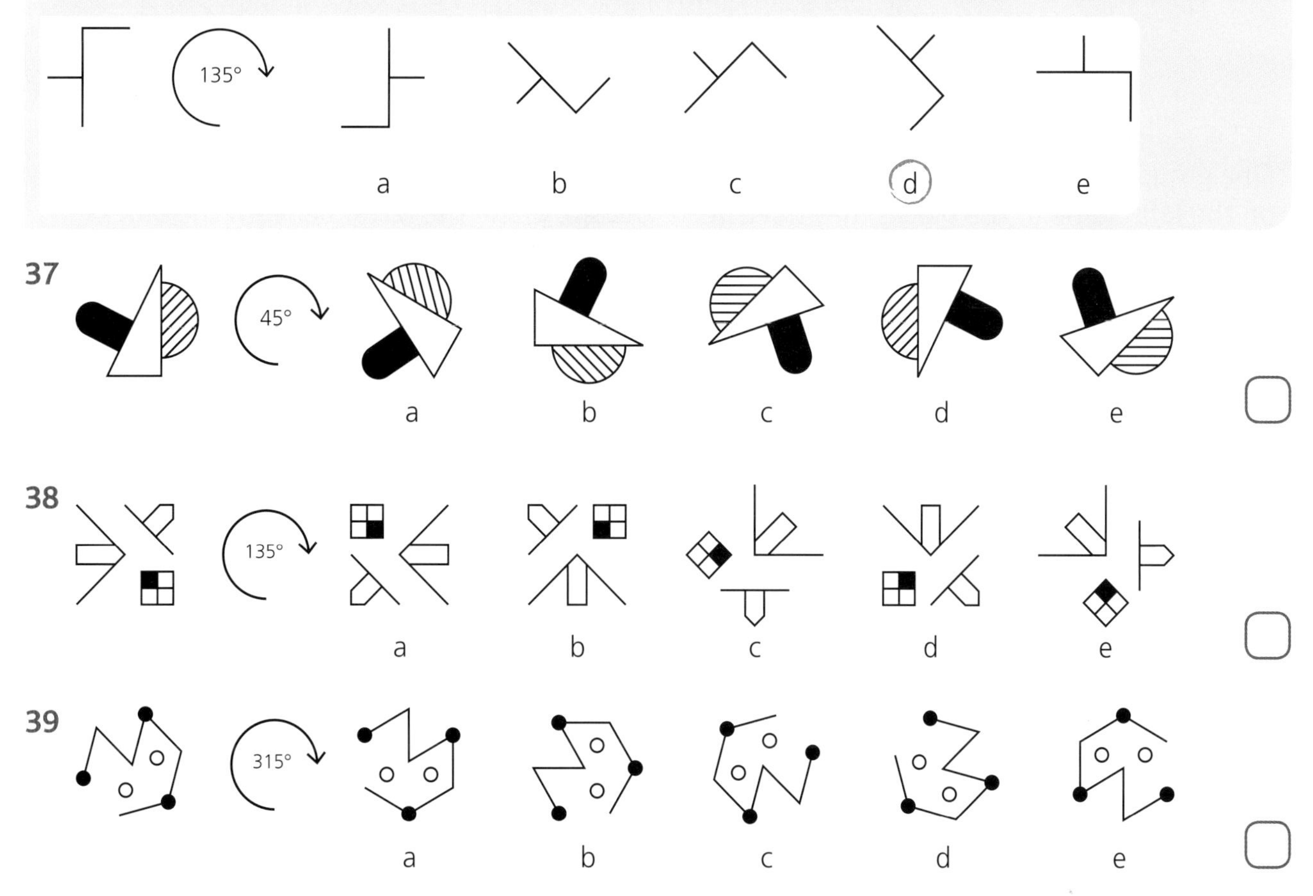

The small shape on the left can be found in one of the pictures in the answer options. It might be made up of one or more pieces. Circle the letter beneath the correct answer. Example:

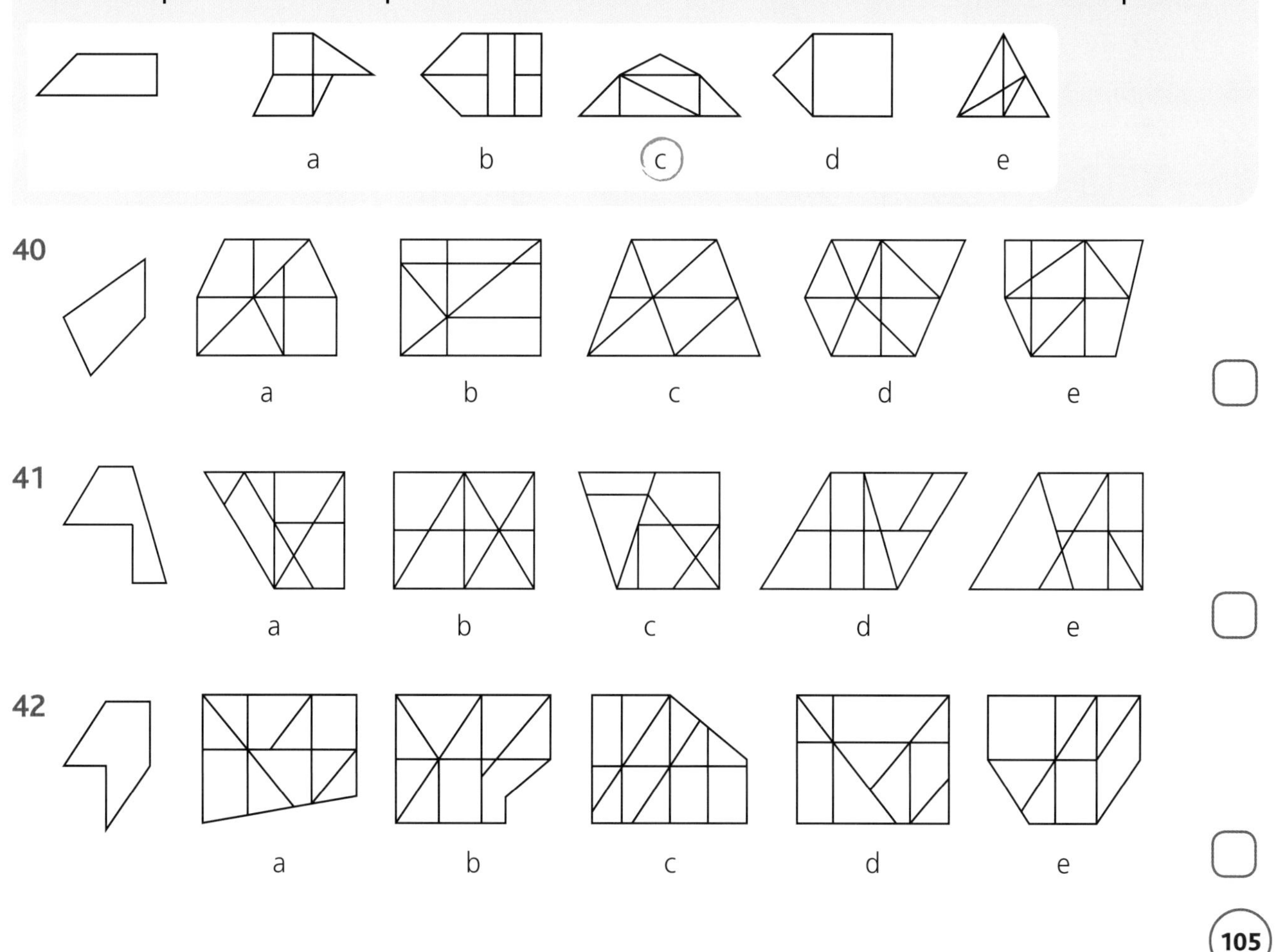

43

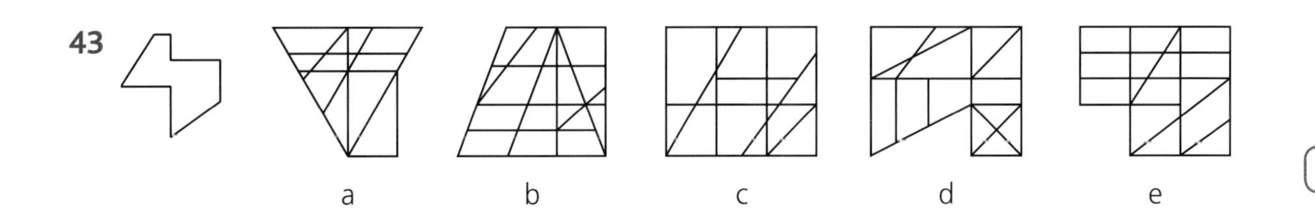

One group of separate blocks has been joined together to make the pattern of blocks shown on the left. Some of the blocks may have been rotated. Circle the letter beneath the blocks that make up the pattern. Example:

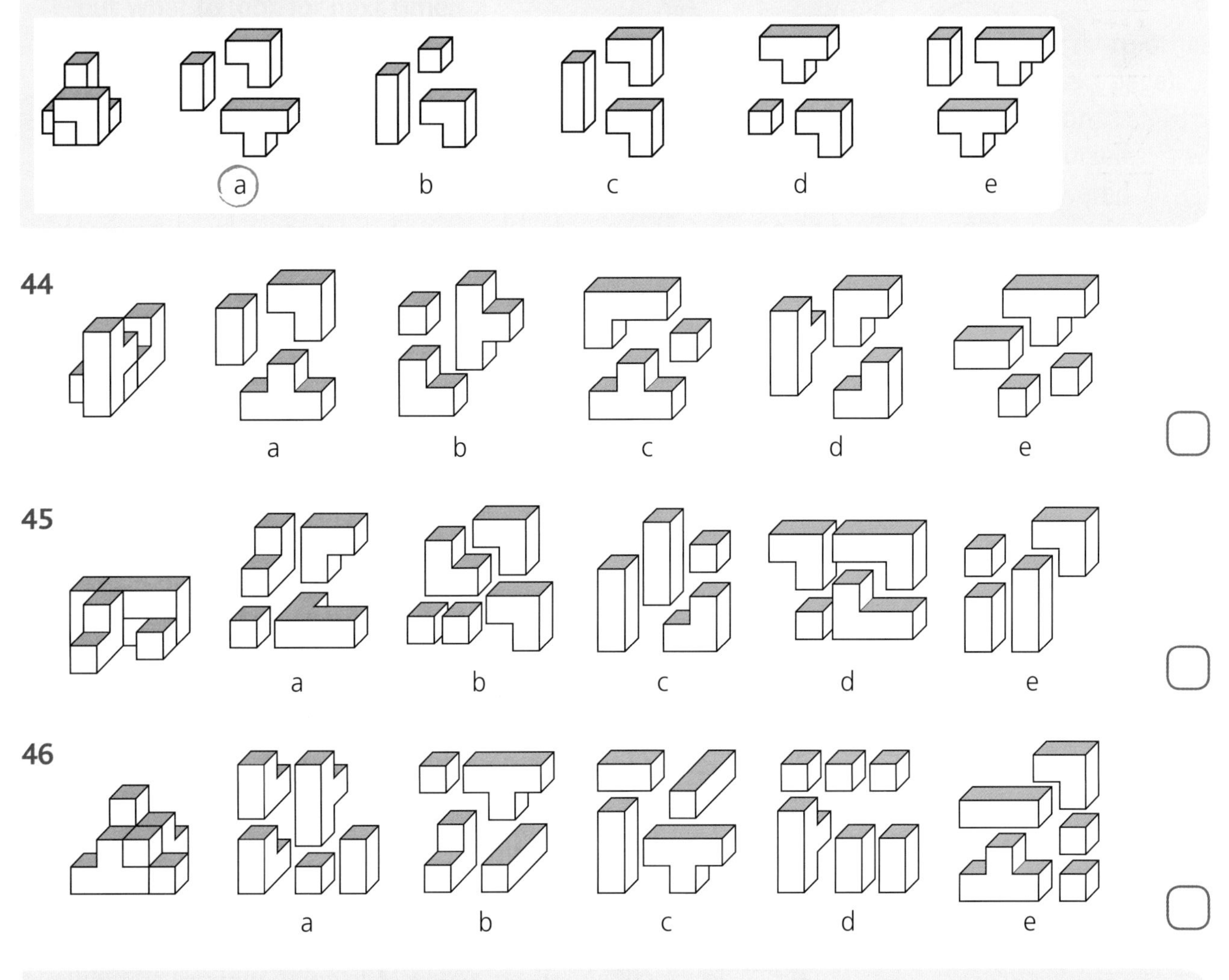

Each letter represents an individual feature in the picture next to it. Work out which feature is represented by each letter. Apply the code to the picture in the box and circle the letter beneath the correct answer code. Example:

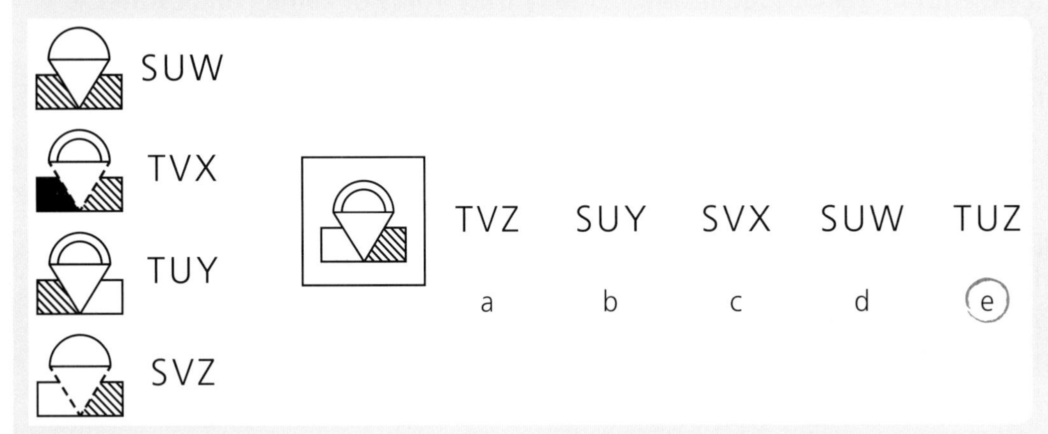

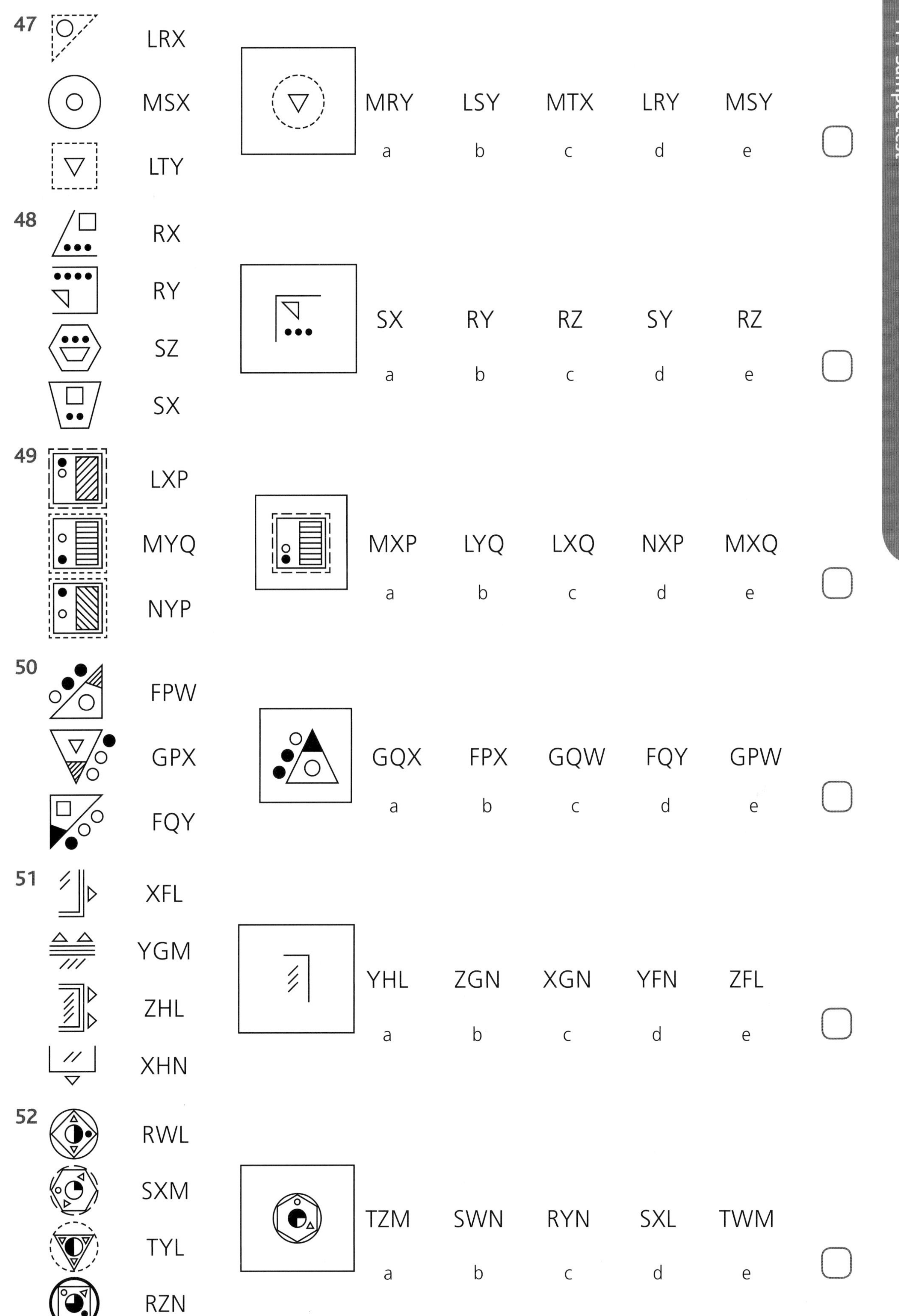
47
LRX
MSX
LTY
MRY a
LSY b
MTX c
LRY d
MSY e
48
RX
RY
SZ
SX
SX a
RY b
RZ c
SY d
RZ e
49
LXP
MYQ
NYP
MXP a
LYQ b
LXQ c
NXP d
MXQ e
50
FPW
GPX
FQY
GQX a
FPX b
GQW c
FQY d
GPW e
51
XFL
YGM
ZHL
XHN
YHL a
ZGN b
XGN c
YFN d
ZFL e
52
RWL
SXM
TYL
RZN
TZM a
SWN b
RYN c
SXL d
TWM e

First you are shown a pattern that is arranged in a sequence. Choose the answer option that completes the sequence when inserted in the blank box. Circle the letter beneath the correct answer. Example:

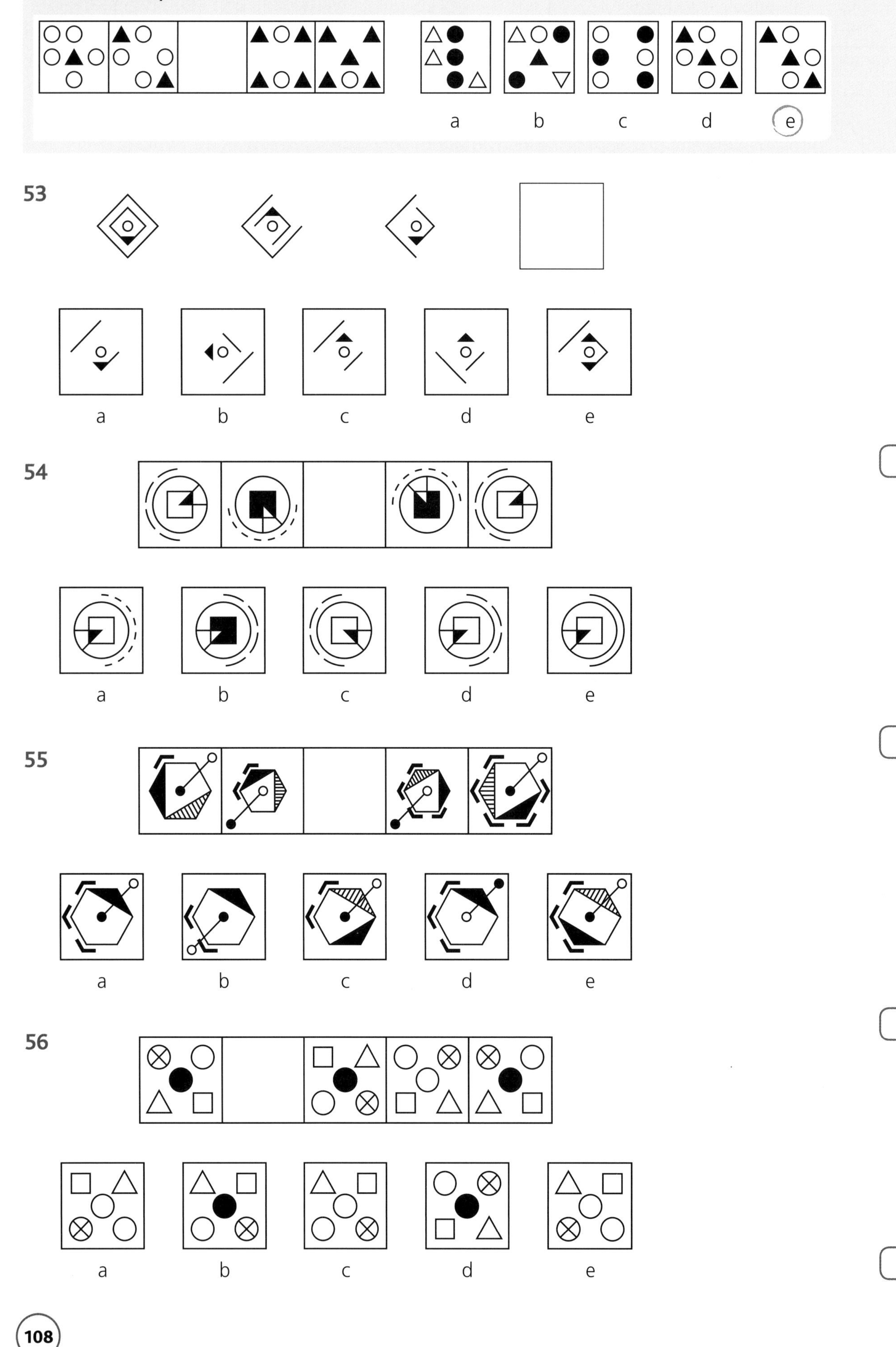

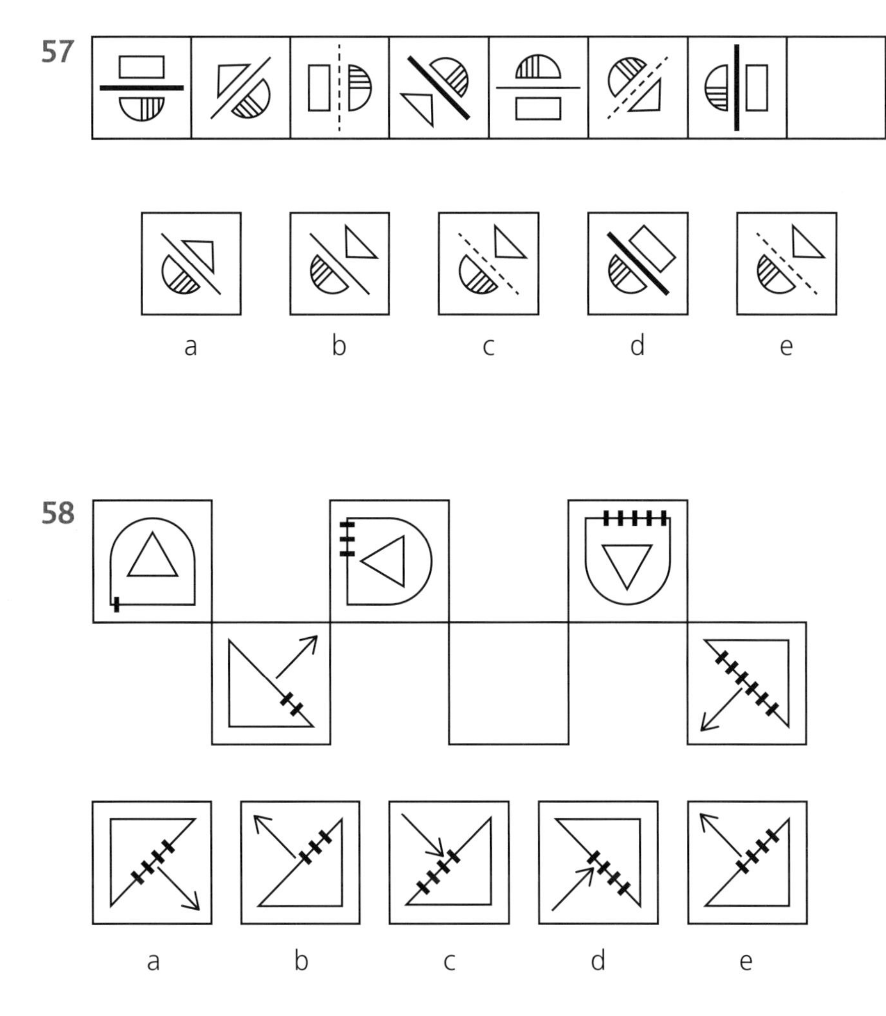

One of the options on the right completes the pattern in the grid on the left. Circle the letter beneath the correct answer. Example:

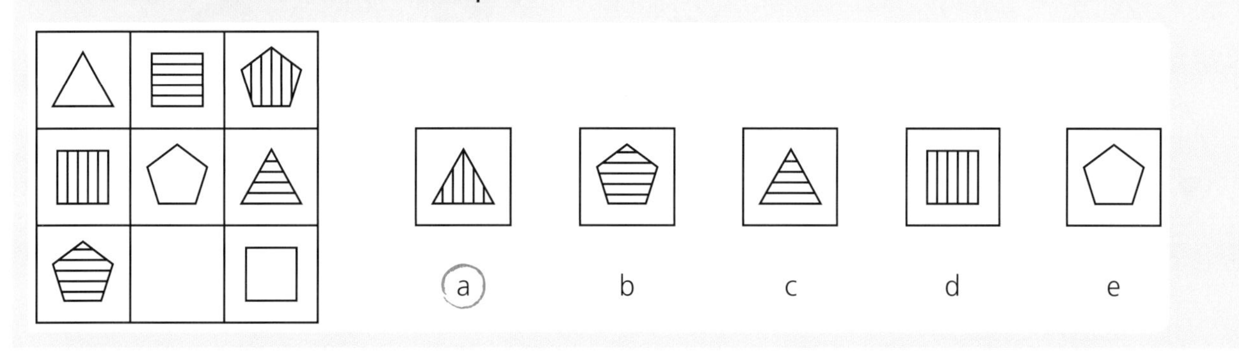

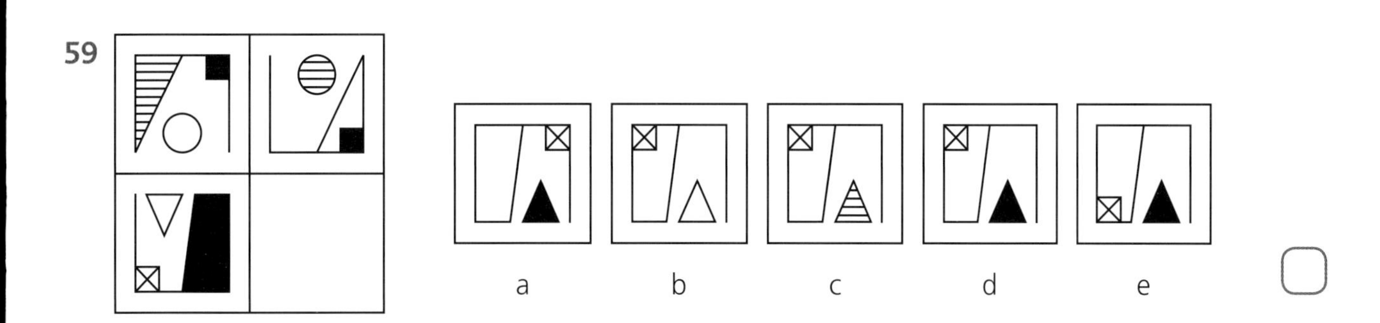

60

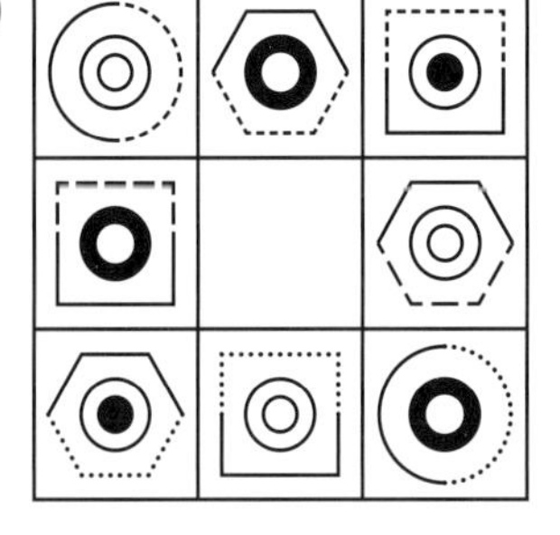

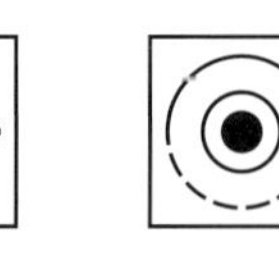 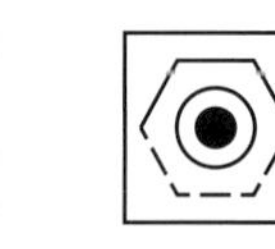 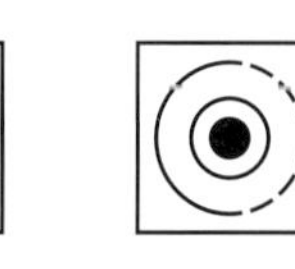 

a b c d e

61

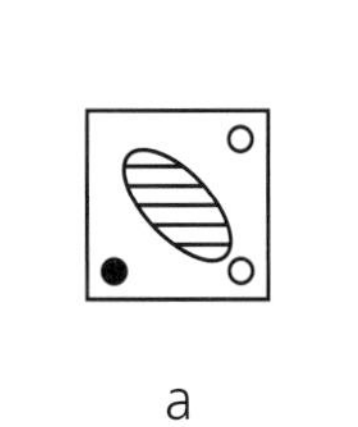

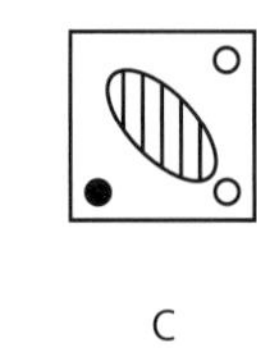 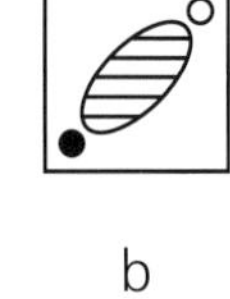 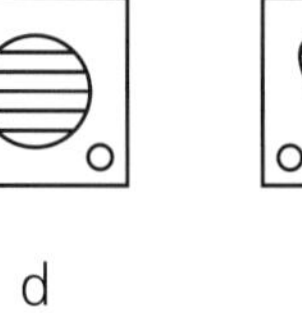 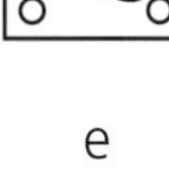

a b c d e

62

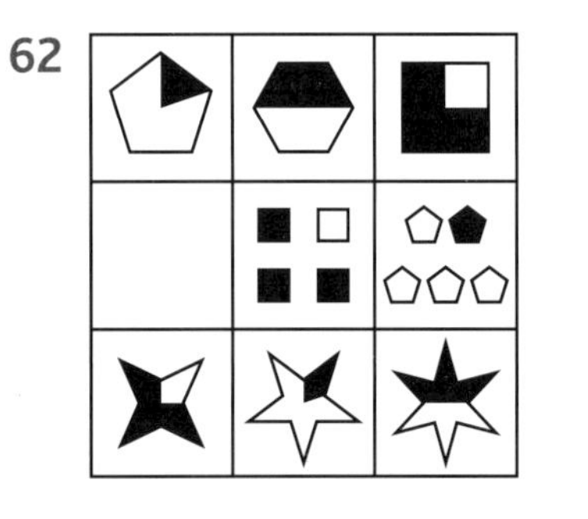

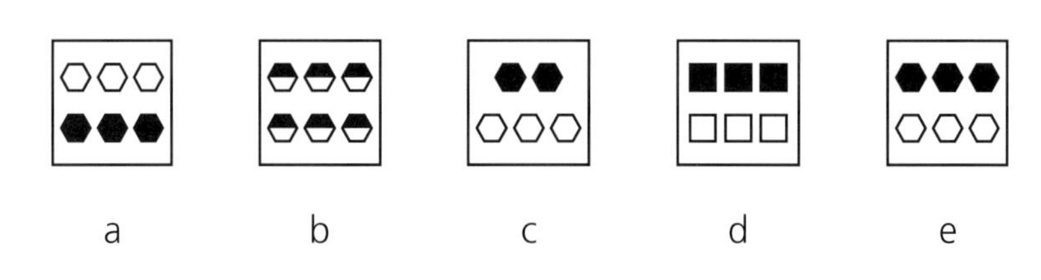

a b c d e

63

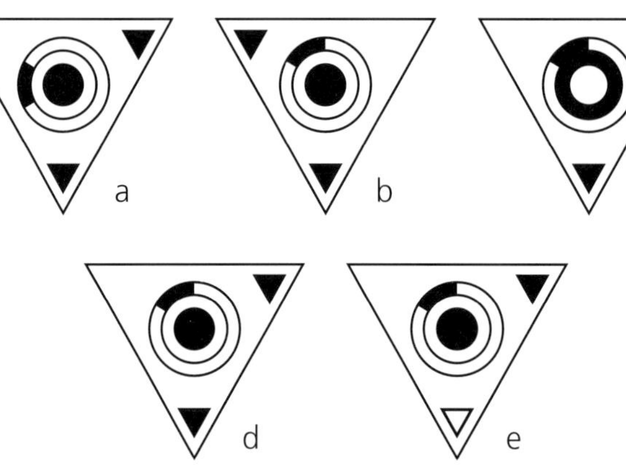

a b c d e

64

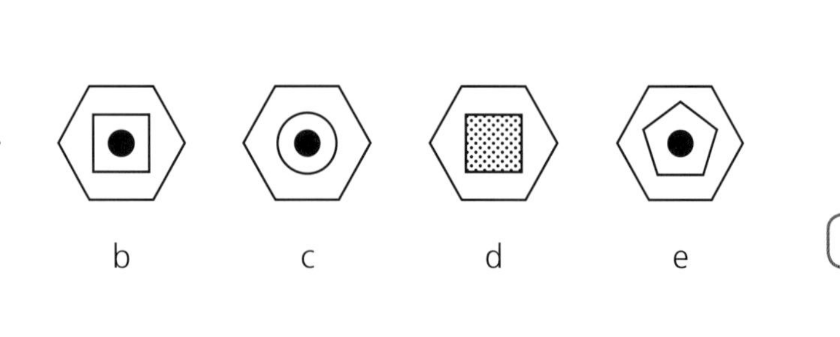

a b c d e

Record your score and time here and at the start of the book.

Score ☐ / 64 Time ☐ : ☐

Answers

Part 1: What is Non-Verbal Reasoning?

1 General themes and question types

What you can expect to see (page 12)

1 (a) Size – the squares get bigger.
  (b) The arrowheads on the solid line change shape.
2 (a) Accept: there are eight images in each box or all the images in each box are the same.
  (b) Accept any two: the shading changes between the boxes, the shapes change between the boxes, the position of the shapes changes between the boxes.

Working out what is important (page 14)

1 (i) the triangles change from white to shaded;
  (ii) the bottom pentagons rotate 180° (or reflect in a horizontal line);
  (iii) the bottom circles change position and overlap the triangles;
  (iv) the small ovals increase in number
2 (a) The pattern of shading in the outer triangles is random.
  (b) Accept any answer that spots a link between the lines in the corner and the position of the shading in the circles: in pictures 1 and 3 there is a long line followed by a short line and the top-left and bottom-right segments of the circles are shaded; in pictures 2 and 4 there is a short line followed by a long line in each corner and the bottom-left and top-right segments of the circles are shaded.

Questions where the picture changes (page 16)

1 All the answers have five lines except for option **c**, which has four.
2 Like the pictures on the left, option **d** is made up of two similar shapes with diagonal shading going from top left to bottom right in the inner shape.

Questions where the picture moves (page 18)

1 a
2 c

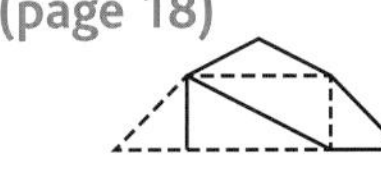

Questions involving algebra (page 20)

1 Accept either of these two options:

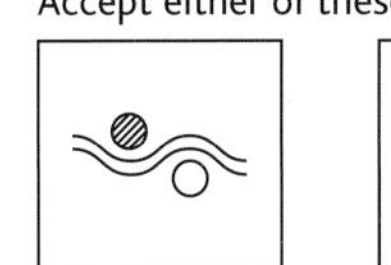

2

Questions involving matrices (page 22)

1

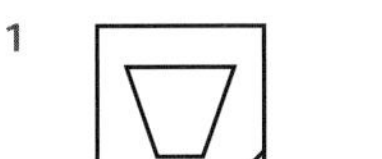

2

Test 1 (page 24)

1 c **number** – all the shapes have five sides except for **c**, which has seven sides
Distractors: **angle** – the angles within the shapes are not important
2 b **shading** – each picture has two black and two white circles except for option **b**, which has three black circles
Distractors: **line style** – of the large circle is not important; **position** – of the elements is not important
3 a **shading** – the pattern in the large shape swaps with that in the inner shape; **reflection** – every element turns upside down (reflects in a horizontal line); **number** – of lines inside the small outer shapes reduces from three to two
4 d **proportion** – the two small shapes at the base increase in size with one fitting inside the other; **shading** – in the large shape becomes the shading in the central shape; **line style** – of the single line at the base becomes the line style of the outer shape
Distractors: **reflection** – the two small shapes in the first picture turn upside down (reflect in a horizontal line) but, as the shapes in the second picture are circles, this is not relevant
5 e **reflection** – the complete picture is reflected in a vertical mirror line so the small semi-circle moves to the left, the 'arrow' at the bottom points to the right instead of pointing to the left and the parallelogram leans to the left
6 d **reflection** – the complete picture is reflected in a vertical mirror line so the central shape leans to the right; **shading** – (a) the black circle remains at the top, (b) the lines in the bottom triangle remain parallel to the central shape, (c) the shading in the top triangle runs parallel to the side that is nearest the black circle
7 a There are two L-shapes in the picture; both are rotated forward 90° and then inverted. One L-shape fits over the single-cube block, the other over the block made up of three cubes in a row. The final two-cube block sits alongside the L-shape at the back.
8 b **shape** – the pictures within a column are identical; **number** – of pictures alternates between rows (3, 2, 3); **rotation** – the shapes rotate 90° anticlockwise, moving down the columns

Part 2: Skills for solving questions

2 Numbers, shapes and relationships

Thinking skills and games (page 27)

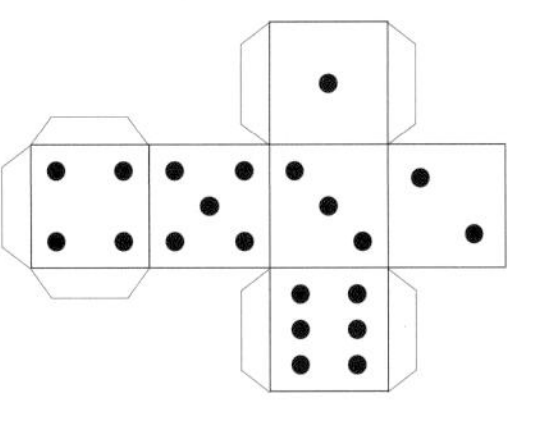

Working with numbers (page 28)

1 (a) The line crosses the edges of the shaded shape three times.
  (b) The number of thin lines in the inner shape equals the number of gaps in the outer shape.
2 c **number** – the large shapes are four-sided
Distractors: **number** – of small circles is not important
3 c **number** – of black circles in each picture matches the number of thick lines on the outer edge of the rectangle
Distractors: **position** – of the black circle and the thick line are not important
4 d **number** – each picture has four small shapes
Distractors: **shape** – the kind of small shape is not important; **number** – of sides in the large shape is not important
5 c **number** – (a) the number of vertical lines matches the number of triangles, (b) the number of horizontal lines matches the number of squares
Distractors: **position** – of the squares and triangles is not important; **shading** – of the shapes is not important
6 b **number** – of white circles matches the number of missing lines, i.e. one line missing – one circle, two lines missing – two circles.
Distractors: **number** – (a) of black circles is not important, (b) of lines on the left is not important, (c) of lines on the right is not important, (d) the total number of circles does not relate to the number of lines

Problems involving scale (page 30)

1 (a)

(b)

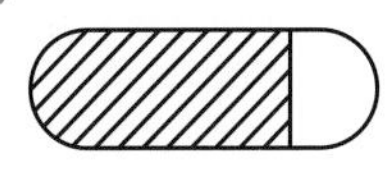

2 d **proportion** – the ratio of large to small shapes is 2:1; **size** – the large shapes become small and the small shapes become large in the second picture
3 e **proportion** – both shapes are cut in half; **shading** – of the two shapes swaps
Distractors: the lower shape rotates to the right 90° but, as there is no answer option that provides the correct shapes and shading in any other orientation, this is a distractor

4 a **proportion/number** – the number of segments in the shape on the left equals the number of sides of the shape on the right; **shading** – the shading of the single segment in the shape on the left becomes the shading of the shape on the right

5 d **scale** – (a) the middle-sized shape (triangle) becomes the largest shape, (b) the largest shape (quadrilateral) becomes a central shape; **proportion** – (a) half of the original central shape is overlapped by the new central shape and is shaded black inside, (b) the original central shape does not change size or shape
Distractors: **rotation** – of the middle-sized shape is a distractor

6 d **proportion** – (a) the horizontal lines running across the vertical line become the sides of the triangle/s, (b) the top line in each set of three becomes the side that comes from the left-hand side of the triangle's base and up to the apex, (c) the middle line is the side that comes from the right-hand side of the triangle's base and up to the apex, (c) the bottom line is the base of the triangle, (d) the proportion of the lines at the top and bottom of the vertical rule swaps

Shapes and shading (page 32)

1 (a) Both stars are made from identical overlapping shapes: the first is a pentagon, the second is a square.
(b) Accept any two identical shapes that form a star, for example:

2 b **shape** – all the others have a line of symmetry that runs through the centre of the smaller shapes too
Distractors: **shape** – (a) whether the shape has curved or straight sides is not important, (b) the central shape is a distractor

3 a **shading** – all the shapes have two parallel sides and the shading runs parallel to these sides, expect in **a** where it is at 90° (perpendicular) to the two parallel sides
Distractors: **shape** – whether regular or irregular is not important

4 b **shape** – all the others have a square base
Distractors: **number** – of sides of the shape is not important

5 e **shape** – the two shapes in each image are identical and create the same shape where they overlap, apart from option **e** where the rectangles form a square
Distractors: **proportion** – the amount of overlap is not important; **shading** – not important; **number** – of sides of the shapes is not important

6 d **shape** – the four large outer shapes made a space in the shape of a square in the centre except in **d** where a rectangle is formed
Distractors: **shading** – of the outer and inner squares is not important; **shape** – the outer shape is not important; **lines** – in the centre are not important

Line styles and angles (page 34)

1 (a) 16
(b) 6
(c)

2 c **line style** – (a) the double line of the pictures always has the thicker line at the top, (b) the line coming out of this double line always bends to the left first

3 d **angle** – (a) the 'flagpole' is always parallel with one side of the shape, (b) the flag always points to the side of the pole that makes the smaller (acute) angle with the intersection with the large shape
Distractors: **shape** – not important

4 e **line style/shape** – the shapes with the dashed lines are regular, that is all sides are of equal length
Distractors: **number** – of sides of the shape is not important

5 a **line style** – at the top of the shield is the same as that of the central square; **angle** – the long bar is parallel to the central line in the square
Distractors: **shading** – of the box is not important; **direction** – of the line at the base of the long bar is not important

6 b **angle/number** – the number of right-angles in the arrow shaft equals the number of tails on the arrow
Distractors: **line** – the number of intersections with the shapes is not important; **shape** – not important

3D views of 2D pictures (page 36)

1

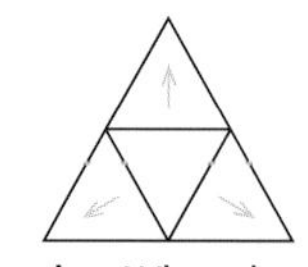

2 d When the net is folded across the three faces that are joined in a horizontal row, the black and white squares are on adjacent faces. Turning the picture to the right and folding in the dotted circle gives the net shown in option **d**.

3 e When the net is folded down the central line and the dotted square is on the top face, the white circle on the white background will be at the front. The face with the square on the right will then fold in to meet these faces on the right-hand side, as shown in option **e**.

4 b When the net is folded across the four faces that are joined in a vertical line, the direction of the diagonal shading on the circle matches the direction of the diagonal bar in the face above it. Folding down the face to the right of the circle leaves the white semi-circle positioned with the straight edge at the opposite edge of the face to the diagonal on the face above it, as shown in option **b**.

5 b When this net is folded up, the triangle is on the face opposite the black arc; the cross is opposite the white diagonal and the vertically shaded diamond is opposite the horizontally shaded segment. Looking at option **b**, when the cross is folded in, it is adjacent to the diamond – the diamond shading being at 90° to the face with the cross. If the three faces joined on a horizontal row are rolled across together, the diagonal then drops down to form the face above. This means that the triangle will form the remaining side next to the diamond, with its base parallel to the shading in the diamond.

2D views of 3D pictures (page 40)

1 kettle **c**; lamp **d**; headphones **a**; cup and saucer **b**

2 b The figure is two cubes wide at the front with a further cube added to the back on the left. The second layer adds an additional cube in height to the cube at the back but does not alter the plan. Therefore the plan is an L-shape like option **b**.

3 d The figure is three cubes wide at the front with a further cube added behind the central cube on the first layer. The two cubes stacked on the central cube at the front do not alter the plan. Therefore the plan is a T-shape as shown in option **d**.

4 d The figure only has one cube at the front, but goes back four cubes in total. The two cubes stacked on the second cube do not alter the plan. However, the two cubes projecting either side create a cross of 3×4 cubes. There are no spaces between the cubes so the answer must be option **d**.

5 a The figure has four cubes at the front (which discounts option **b**). It is three cubes deep since the taller stacks are clearly sitting on a cube (discounting options **d** and **c**). There is a cube between the two stacks at the back and further shading in front shows that there is a space, so the answer must be option **a**.

Test 2 (page 42)

1 c **shading** – the direction of the shading is the same in both shapes of each option except for **c** where the stripes go in different directions
Distractors: **position** – of the triangles is not important

2 e **shape** – the triangle overlaps a curved side, except for in **e** where it overlaps a straight side
Distractors: **shape** – (a) of the triangle is not important, (b) of the second shape is not important; **line style** – of the large shape is not important

3 d **number** – all the shapes with straight edges have six sides except for option **d**, which has four
Distractors: **proportion** – the size of the shapes not important

4 a **proportion** – the top shape reduces and the bottom shape enlarges; **shading** – the large shape becomes black but the small shape keeps its original shading
Distractors: **position** – the shapes rotate 45° but all the answer options show the correct rotation

5 c **number** – of sides in the large shape halves; **line style** – (a) the outer shape retains its line style, (b) inner line style is taken from the middle shape; **shading** – comes from the small shape in the centre

6 c **line style** – all the shapes have at least one dashed line; **shape** – all the shapes are symmetrical and are divided in half
Distractors: **shape** – the number of straight and curved sides is not important; **number** – of dashed lines is not important

7 b **proportion** – each figure has two lines the same length and one different; **shading** – there is always a black and a white circle at the end of the outer lines
Distractors: **angle** – the angles made by the three lines are not important

8 e **number** – (a) there are always three lines, (b) there are always three circles; **proportion** – the three lines are of equal length
Distractors: **position** – of the elements is not important; **shading** – of the circles is not important

9 d **shape** – (a) each figure is made up from a circle and another shape with no straight edges, (b) the entire shape has one line of symmetry
Distractors: **position** – of the circle is not important; **number** – of curves of the shapes is not important

10 a **proportion** – the square is split into four equal parts; **shading** – (a) the shading is always white, black, diagonal and dotted, (b) the black and dotted shading are always next to each other
Distractors: **proportion** – the size of the outer square is not important

11 a When the net is folded the face with the arrow will be pointing away from the right-hand side of the cross. This will fold down to touch the 'U'-shape, the base of which joins the bottom of the face with the cross (opposite the black triangle).

12 c When the net is folded the face with the open top of the 'U'-shape meets the cross. The right-hand side of the 'U' is joined to the three circles, which run at 90° to it.

13 b When the net is folded the faces with the circle and the square are next to each other (as they are on the central line). The base of the two black triangles folds down onto the cross (think of the green arrows in the 'elephant' diagram), meaning the circle will then fold in to complete the net.

14 d Neither **a** nor **b** are nets of a cube. In **c** the shapes will run in a line around the cube so the hexagon and circle will not be next to each other. In **e** the hexagon will not be next to the circle.

15 e Neither **a** nor **b** are nets of a cube. In **c** the triangle and circle will be on opposite sides, while in **d** the angle of the rectangle is wrong when folded to meet the side with the triangle.

16 b The bottom layer of the stack of cubes has three cubes at the front, so the bottom of the 2D plan must have three squares in a row at the bottom. There is then a gap to the next stack of cubes where there is a single cube to the front, so the plan will show a space and then a single square. The back of the stack has three faces showing in a row, so the top of the plan will have a row of three squares, as in plan **b**.

17 a The front stack of cubes has three in the front row and two in the second (on two levels) so the plan will show two squares with a further three below. The second stack of cubes only has two faces showing, so the answer must be **a**.

18 e The front stack of cubes has three in the front row and a further cube (on the second level) on the next row so the plan will show one square with a further three below. The next row has two cubes (on two different levels) with the back row being made up of three cubes, so the answer must be **e**.

19 d There are two cubes on the front row with no overlapping cubes from the levels above. There are two cubes in the next row with a gap before the stack of cubes at the back. There are four faces showing on the stack at the back, which sit next to each other and also line up with the second row. So, there must be a block of six squares in the plan. Therefore, the answer must be **d**.

20 c The first two rows have a single cube. The following row has a single cube on the bottom level with two cubes on the right forming a row of three squares on the plan. There are three faces showing on the back row, so there will be another row of three squares on the plan. The answer must be **c**.

3 Links of position and direction

Placement (page 48)

1 (a) **9**: there are four small white squares, four small black squares and one large transparent square
(b) **12**: the large triangle is made up of three triangles, with an additional triangle in the middle. These four plus the two black triangles make six. There is also a medium-sized white triangle above the horizontal line running across the triangle, four small triangles and one inverted white triangle.

2 b **position** – the two rows of shapes swap position so that the top row becomes the bottom row and vice versa
Distractors: **rotation** – the shapes do not rotate, although a number of the incorrect answer options do

3 c **position** – (a) the two shapes on the left move one shape to the right, (b) the right-hand shape moves to the position of the first shape

4 b **position** – (a) the top and bottom sides of the square swap places, (b) the left and right sides of the square also swap places, (c) the 'arrowheads' move one line to the right
Distractors: **number** – of shapes on the arrow tails do not change

5 e **position** – (a) the shapes on the left weave 'over, under, over, under' (moving left to right) and on the second picture change to the opposite pattern 'under, over, under, over', (b) the shape on the right weaves 'under, over, over, over' (moving right to left) and in the second picture the pattern changes, to almost the opposite with the shape on the right weaving 'over, under, over, under'; **shading** – the shading of the shapes moves: top to middle, middle to bottom, bottom to top

Symmetry and vertical reflections (page 50)

1 (a) a cross or 'X'
(b) two

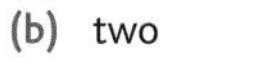

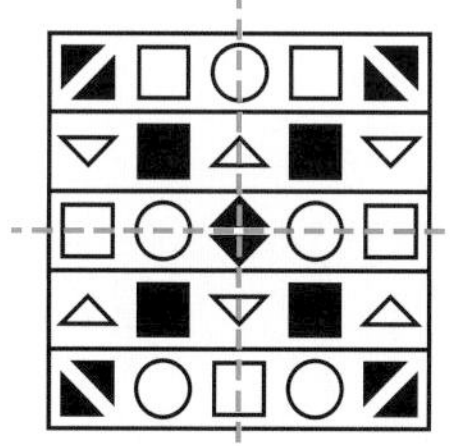

2 c When reflected vertically, the rectangle moves from top left to top right, the black 'signal' shape moves from middle right to middle left and the triangle moves from bottom right to bottom left. Each shape and the cross pattern all change to the opposite shading.

3 d When reflected vertically, the triangle moves from top left to top right, the pointed flag shape moves from middle right to middle left and the rectangle moves from bottom left to bottom right. Each shape and pattern changes to the opposite shading.

4 b **reflection** – the complete picture is reflected in a vertical mirror line so the 'step' shape is on the left, rather than on the right of the image and the diagonal on the square runs top left to bottom right
Distractors: **position** – in option **d** the only difference is that the diagonal is shown in the wrong position; **reflection** – the picture has been reflected in a horizontal mirror line in option **e**

5 a **reflection** – the complete picture is reflected in a vertical mirror line so the whole image will be pointing to the right rather than the left
Distractors: **rotation** – in **d** the outer triangles are rotated; **reflection** – the picture has been reflected in a horizontal mirror line in option **e**

6 c **reflection** – when the complete picture is reflected in a vertical mirror line, the largest L-shape on the right will point to the left and be on the left of the picture
Distractors: **position** – option **b** has random rotations from the original picture and **d** is reflected vertically with a number of random rotations; **reflection** – **a** is reflected horizontally, as is **e** with additional rotations

More reflections (page 52)

1

2 c The fold works like a horizontal reflection or mirror line so there will be two sets of three circles. The central circle will be at the bottom of the sheet with the two outer circles nearer the line.

3 e The diagonal fold works like a mirror line. Drawing lines from the vertices to the line shows that the 'V'-shape inverts so that it is pointing at the line, from the same distance on the other side.

4 a There are three layers in the final folded diagram, which shows the top of these three layers. This means that the right-hand side of the rectangle is the mirror line since this was the last fold. The image reflects in this line so it flips left to right with the top circle being on the right. The first fold was from the left, so this is the new line of reflection. As the bottom layer will be the same as the pattern shown here, the holes on the left will look exactly the same as the holes on the right.

5 d The final fold goes through four layers since the first fold stops at the bottom of this triangle. When the triangle reflects along the diagonal line, a circle is also seen in the top corner. When the second (vertical) fold is reflected, both circles are reflected and the image flips left to right. The arrows also reflect and are both the same way up. The first fold is a horizontal reflection so there will be four arrows upside down in a row.

Rotations with 2D pictures (page 54)

1

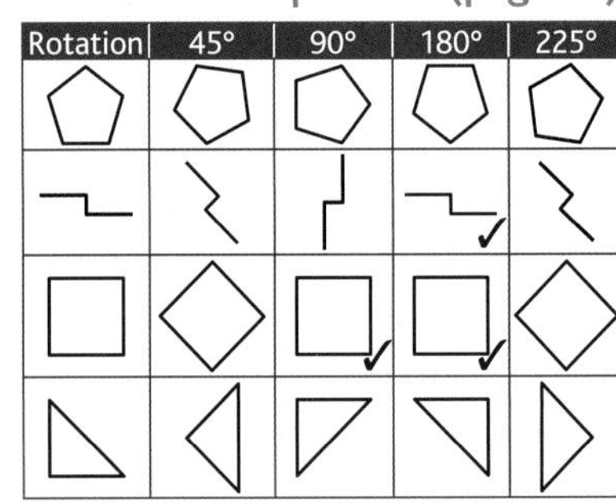

2 e The picture has been rotated so that it is appears upside down in option **e**, so it has been rotated through 180°. The other options have been rotated through the following degrees: **a** 225°, **b** 90°, **c** 45°, **d** 135°.

3 b 135° is one and a half right-angles. The other options have been rotated through the following degrees: **a** 225°, **c** 45°, **d** 180°, **e** 90°.

4 c **rotation** – the picture is rotated 180°; **position** – the arrow is pointing to the white triangle; **proportion** – of both the arrow and shading match the original picture

5 e **rotation** – the picture is rotated 270° clockwise or 90° anticlockwise; **position** – (a) the white arrow shape forms the top layer followed by the black and tinted shapes, (b) in relation to other shapes, the white arrow shape points away from the rest of the picture and one vertex of the hexagon touches the long edge of the rectangle shape; **shape** – the shapes are the same: an irregular pentagon with two right-angles, a hexagon and a rectangle with the corners cut out

6 b **rotation** – the picture is rotated 210°; **position** – (a) the small hexagon has a side parallel to the rectangle, (b) the triangle sits on the arrow
Distractors: **reflection** – answer **c** is a reflection, as you can see from the diagonal shading

Rotations with 3D pictures (page 56)

1 (a) 90° or 270° (b) 180°

2 e The line from the front to the back of the top face on the first cube has moved to the left face of the second cube. The triangle on the front face has rotated 90° anticlockwise to point to this line, so the whole cube must have been rotated 90° anticlockwise. Rotating the third cube in the same way will have no effect on the circle on the front face. The lines on the top face move to the left-hand face, with the longer line going from the bottom rear corner to the top front and the shorter line going from the top rear corner to the middle.

3 b The line from the front to the back on the left face on the first cube has moved to the front face of the second cube. The square on the top face appears unchanged, so the whole cube must have been rotated 90° clockwise. Rotating the third cube in the same way will rotate the line on the top face so that it runs across the face rather than down. The black circle on the left face will move to the front and the white square will move out of sight.

4 b Shape **b** is rotated 45° away from you around a vertical line and then rotated forward 90° around a horizontal line, so that it stands on the single and double block.

5 c Shape **c** is rotated 90° clockwise as if it is spinning around on its back, and then rotated 90° around a horizontal line so that it looks as if it is moving from lying down to standing up. This means that the single cube is at the back and cannot be seen.

6 a Shape **a** is rotated 180° clockwise as if it is spinning around on its back, and then rotated 135° around a horizontal line.

Translations with 2D pictures (page 58)

1 a cat

2 d **position** – (a) the circle is inside the square with cut-out edges so is not visible, (b) two points of the star stick out from the left side of the square, (c) one vertex of the quadrilateral sticks out to the right

3 b **position** – (a) there is a small gap between the right-hand side of the arc and the circle, (b) the arc leaves a space along the right of the large shape, (c) the star only sticks out on two sides
Distractors: **proportion** – the star does not change in size

4 c The shape is sticking out to the right of the square in **c**. The parallelogram in **a** is too flat, the shape is rotated in **b**, in **d** it is reflected vertically and in **e** it is reflected vertically and reduced in size.

5 e All the vertices of the shape touch the edges of the outer box in **e**. Option **a** is a similar shape but the angle formed on the top right is a right-angle and the middle point is nearer to the 'base'. Option **b** is rotated with the shape noticeably distorted. Option **c** is rotated with no change to shape.

6 d Follow the right-hand side of the top trapezium. The shape runs across two shapes from the bottom corner of this shape and ends on the line below.

Translations and rotations (page 60)

1 four

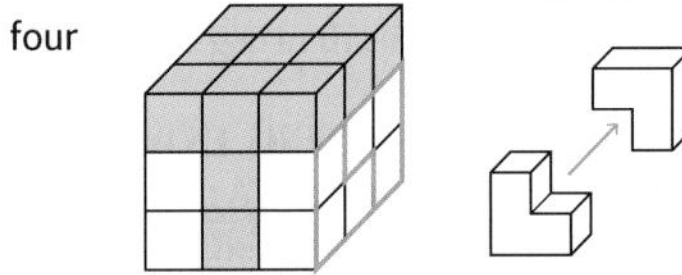

2 c, d, e Option **c** rotates 90° to form the right-hand side section at the base of the shape, **d** does not rotate and forms the left-hand side of the shape, **e** rotates 45° clockwise to sit on top of **c**.

3 a, c, e Option **c** does not rotate and forms the bottom left-hand side of the shape, **e** rotates 135° clockwise to form the right-hand side of the shape, **a** rotates 180° to sit on top of **c**.

4 b One block is rotated 90° clockwise around a vertical line – this means that it spins around on its base so that the single block is at the bottom front. The second block is rotated 180°, around a horizontal line so that it is an upside-down L and fits over the first block.

5 d One T-shape faces forward, while the other lies down so that the long edge is on the right. These two fit together. Then the two-cube block is placed on the bar of the T that is lying down.

6 a The long L-shape rotates around a horizontal line 90° so that the short part of the L is pointing forwards and it is lying on its long edge. The shorter L-shape fits against the base of the large L. The T-shape rotates 180° around a horizontal line so that it is upside down and fits over the back of the long L-shape.

Test 3 (page 64)

1 c **angle** – two lines join at right-angles to form a square with a corner of the large shape, overlapping it; **shape** – (a) two lines cross to touch the corners of the small square, (b) each picture includes a circle
Distractors: **line style** – the thickness of the circle line is not important

2 b **shape** – (a) the small overlapping shape is the same as the larger shape, (b) the separate black shape is the same as the outside half of the small overlapping shape; **position** – half the smaller shape overlaps the larger shape; **shading** – the inner portion of the overlapped shape is shaded black, as is the small shape inside
Distractors: shape – the type of shape is not important; **position** – of the shapes around the edge of the larger shape is not important

3 e **direction** – the larger shape always points to the right; **number** – there are always only five points of the star showing
Distractors: **shape** – (a) the shape of the large image is not important, (b) the type of star is not important; **position** – whether the star is in front or behind the larger shape is not important; **line style** – not important

4 a, b, e **a** and **e** sit one above the other on the left, **b** sits on the right-hand side

5 b, d, e **d** rotates 90° clockwise and sits across the top of the shape, **e** rotates 180° to form the left-hand side of the shape, **b** forms the square to the right of the centre

6 a, c, e **e** sits bottom left, **c** rotates 45° clockwise and then sits next to the right of **e** to form the shape at the bottom right, **a** rotates 180° to form the top right of the shape

7 c The fold lines act like lines of reflection. So, when unfolding the second fold (a vertical reflection), there will be two circles sitting next to each other (in a horizontal line) on the right-hand side. These three circles are then reflected in the diagonal line when the final fold is opened out.

8 a The fold lines act like lines of reflection. So, when the second fold is opened out (a diagonal reflection), the heart reflects diagonally and points inwards. The two hearts and one circle then reflect in a vertical line when the final fold is opened out to give the final pattern.

9 b The fold lines act like lines of reflection. So, when unfolding the second fold (a diagonal reflection), there will be a second arrow that also points upwards to the corner as well as a heart pointing upwards on the top side and a circle on the left-hand side. When the final fold is opened out, the arrows reflect in a diagonal line with all the arrows pointing inwards.

10 c 45° is half a right-angle. Rotating the shape clockwise by 45° gives option **c**.

11 a 225° is two and a half right-angles, so the shape should be pointing in the opposite direction and then turned a further 45°. Check option **a** is the answer by looking at the position of the diamond and triangle.

12 d 135° is one and a half right-angles, so option c would be a quarter turn and option **d** is another 45°.

13 c Shape **c** is rotated 90° backwards and then 90° clockwise around a vertical line.

14 d Shape **d** is rotated 90° clockwise around a vertical line.

15 a Shape **a** is rotated 90° clockwise around a horizontal line.

16 b Shape **b** is rotated 90° anticlockwise around a horizontal line and then rolled forward 90°.

17 a The T-shaped block is turned upside down and rotated through 90° to sit over the block that is three cubes long. The L-shape then fits over the other end with the single block in the middle.

18 e The L-shaped block rotates 180° to lie flat. The single block then goes on the corner and the block that is two cubes long sits on the right-hand side. The final three-cube block sits across the top.

19 d The T-shaped block rotates 90° forwards and then turns to face the front. This block fits over the two-cube block and the single cube then sits alongside at the base. Finally the two L-shapes form the shape on the left.

20 c The long L-shaped block rotates 180° and then drops forward to lie flat. A single block sits on top of the L at the back. The two-cube block fits at the front and the short L sits over the other end of the long L on the left. The final single-cube block sits on the top on the left of the short inverted L.

Part 3: Solving complex questions

4 Connections with codes

Codes using numbers and shapes (page 70)

1 (a) end (b) fob

2 d **number** – the letter D represents two small horizontal lines and the letter E represents three lines; **shading** – M represents diagonal, N white and O black shading

3 c **scale** – the letter J represents a large cross and K a small cross; **line style** – the letter P represents a thin solid line, Q a thick solid line and R a dashed line

4 a **proportion** – the letters D and E represent the number of small to large circles (D is one small to two large, while E is two small to one large); **shape** – M represents a cube and N a cylinder; **shading** – V represents one shaded circle and W represents two shaded circles

5 b **shading** – the letter A represents a black triangle, B a triangle with horizontal shading and C a triangle with vertical shading; **angle** – S represents a narrow angle at the top of the 'Y' and T stands for a wider angle; **number** – X represents two circles while Y represents four circles
Distractors: **position** – of the shaded triangle is not important

Codes using position and direction (page 72)

1 The arrow is white and does not touch the outside box.

2 c **position** – (a) D means that the star is outside the L-shape while E means the star is inside it, (b) G represents an upright L-shape, H means the bottom corner is at the top right and I shows the corner top left

3 b **vertical reflection** – the letters A and B stand for whether the Z-shape is reflected in a vertical line so A is reflected and B is not; **direction** – the letters N, O and P represent the direction of the missing quadrant in the circle so with N the gap points top right, O the gap points downwards and P the gap points top left

4 c **direction** – the letter D represents the triangle pointing upwards, E pointing downwards, F pointing to the left; **position** – (a) U stands for the top and bottom petals overlapping and V for the left and right petals overlapping, (b) X represents the triangle not touching the circle while Y is touching the circle

5 a **position** – (a) the overlap of the two squares forms a shape so M stands for a square shape and N for a rectangular shape, (b) W, X and Y represent the position of the crosses so W is diagonal from top left, X is vertical and Y is diagonal from bottom left
Distractors: **shading** – not important; **scale** – size of the squares is not important

Codes combining different skills (page 74)

1 Illustration of a larger or smaller circle with vertical shading: A stands for black shading and B for vertical stripes, while the second letter of the code stands for the size of circle and so the circle drawn can be any size other than the ones already shown.

2 d **line style** – the letter D represents a dashed line, E a solid line and F a dotted line; **reflection** – the letter R shows the central arc reflected and in S all the arcs are the same way up (it does not matter which way up)

3 c **shading** – R represents the order of shading (top to bottom) black, white dotted and S stands for the order white, dotted, black; **position** – the order of the layers is represented by the letters M, N, O, with M showing the bottom triangle at the front, N the top and bottom triangles at the front and O the middle triangle at the front

4 b **position** – the letter A represents the small circle at the point of the arrow on the left while B is on the right; **proportion** – of the two arrows shaded (when added together) is represented by letters R and S, with R standing for one whole arrow shaded and S for one and a half arrows shaded; **direction** – W represents left arrow up and right down, X both up, Y left down and right up

5 e **angle** – (a) A represents an obtuse angle made between the long line and the short line at the top right while B represents an acute angle, (b) E represents an obtuse angle between the long line and the short line at the bottom left with F being a right-angle and G an acute angle; **position/line style** – the position of the thick line and the line style of the outer box are represented by T, U and V so T equals a dotted box outline and a thick top line, U is a solid box outline and a thick bottom line and V is a solid box outline and a thick top line.
Distractors: **shading** – the dotted shading is not important

Test 4 (page 76)

1 a **shape** – (a) the upper code represents the large shape with R for lozenge, S for rectangle and T for diamond, (b) the bottom code represents the small shape with G for circle and H for triangle

2 c **direction** – the top code represents the direction of the arrow with F for an arrow pointing right and G for an arrow pointing up; **line style** – the bottom code represents the style of line the arrow is drawn in with X for a thin solid line, Y for a thick solid line and Z for dashed
Distractors: **position** – of the small circle is not important

3 b **proportion** – (a) the top code represents the size of the central circle with R for small, S for large and T for medium, (b) the bottom code is for the proportion of this shape that is shaded with X for half, Y for one-quarter and Z for three-quarters
Distractors: **shading** – of the small circle is not important

4 d **position** – the top code represents the position of the rectangle so in P it is at the top, Q in the centre and R at the bottom; **shading** – the top code represents the shading of the rectangle with K for black, L for vertical stripes and M for white
Distractors: **number** – of triangles is not important; **position** – of triangles is not important

5 d **shading** – the top code represents the shading of the top left corner of the box where V is striped, W is black and X is white; **number** – the bottom code represents the number of circles with L for three, M for two and Z for four
Distractors: **shading** – of the lower triangle is not important

6 c **shape** – the letter S represents a square, T a triangle and V a flag; **shading** – K represents diagonal shading, L a dotted pattern, M white and N black
Distractors: **number** – of small circles is not important

7 b **shape** – the letter F represents the square, G the 'arrow' shape and H the L-shape; **number** – R stands for one small triangle and S for two
Distractors: **number** – of circles is not important; **position** – of the triangles and circles is not important

8 b **direction** – (a) the first letter represents the direction the lower triangle is pointing in with F to the right and G to the left, (b) the second letter represents the direction the upper triangle is pointing in with S to the left and T to the right; **shading** – the third letter represents the shading with X meaning one triangle black and Y no shading

9 e **direction** – the first letter represents the direction of the V-shape on the right so R points to the right, S to the left and T upwards; **shape** – the second letter represents the large shape with X meaning a square, Y a circle and Z a shield; **proportion** – the third letter represents the amount of the large shape that is shaded with K standing for half, L for less than half, M for more than half and N for the whole shape
Distractors: **shape** – the small circles are distractors but are difficult to discount initially as they could also be the solution to the first letter. However, there is no answer option that allows RZN as a solution.

10 a **number** – (a) the first letter represents the number of small diamonds with F standing for two, G for three and H for one, (b) the second letter represents the number of circles with S for three, T for one and V for two, (c) the third letter represents the number of upside-down V-shapes with X for three, Y for two and Z for four
Distractors: **shading** – not important

5 Sequences

Sequences using numbers and shapes (page 80)

1 Three extra pictures to be added. Answers may vary. For example:

2 a This is a reflecting and alternating sequence: **shading** – the semi-circles alternate between black and white; **scale** – the rectangles at the base of the 'mushrooms' reduce in size towards the centre, and so disappear in the central position

3 d **number** – (a) moving left to right, one extra square is added at the base, (b) three extra lines are added to the cube pattern in each box; **shading** – of the squares at the base alternates between black and white, with the square on the left always remaining white

4 e **proportion** – moving left to right, the large shape becomes the smallest, the smallest becomes the medium-sized shape and the medium-sized shape becomes the largest; **line style** – the styles alternate between the boxes

5 b **number** – (a) the large shape increases by one side each time, (b) the number of zig-zags on the line equals the number of sizes in the large shape; **shading** – the background alternates between black and white; **angle** – each shape contains a right-angle

Sequences using position and direction (page 82)

1 The sequence should show three further boxes before the frogs meet.

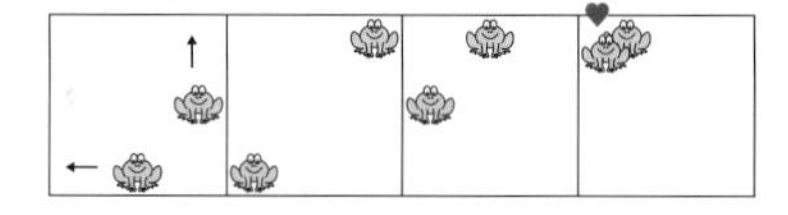

2 a **rotation** – (a) the cross shape rotates 90° (the direction does not affect the image), (b) the circle rotates one division clockwise
Distractors: **line style** – this does not reflect

3 b **position** – (a) the bullet shape moves up one-third of the box each time and then repeats, (b) the bullet shape becomes black where it overlaps the rectangle, (c) the clear circle alternates in front of and behind the rectangle

4 c **reflection** – the L-shape on the left reflects in a horizontal line and the L-shape on the right reflects in a vertical line in every alternate square; **position** – the L-shapes move down one-third of a square every two squares; **number** – one semi-circle is added to the opposite side of the line each time at each end; **shading** – existing semi-circles turn black, new ones are white

5 e **position** – the shading on the triangle swaps position in alternate boxes; **rotation** – (a) the gap in the outer circle rotates clockwise 45°, (b) the gap in the middle circle rotates anticlockwise 45°, (c) the gap in the inner circle rotates clockwise 45°

Sequences combining different skills (page 84)

1 (a) **box 21**: A segment is lost every two boxes (the lower boxes retain the same number of segments as the previous upper box). This means that an orange segment will disappear in the next box along (box 5). A green segment will be lost next so another orange segment will disappear after another four boxes, and so on. So, orange segments disappear in boxes 9, 13, 17 and 21.
(b) **box 24**: The blue snake loses its segments after the orange snake, so it will be the last to disappear. If the last orange segment goes in box 21, the head of the top snake will disappear in box 23, so the blue snake will disappear in box 24.

2 a **number** – of lines increases by one moving left to right, and so decreases by one from right to left; **shading** – changes every third box; **rotation** – the shading moves around one star point in a clockwise direction, moving left to right, so moves anticlockwise from right to left

3 d **proportion** – the 3D shape reduces by one-quarter between boxes; **shape** – the shapes in the bottom box are a 2D view of the front face of the previous 3D shape; **shading** – of the small shape matches the background of the previous 3D shape

4 e **position** – moving left to right, the line with a small shape on it on the right-hand side moves left and all the other lines move along too; **shape** – the large shape matches the small shape at the end of the line on the right; **line style** – of the large shape changes every third box

5 b **number** – moving left to right, there is a repeating sequence on the second row with one less square with diagonal shading in each box (down to no diagonal striped squares) and then the sequence begins again with three shaded squares; **shading** – the top and bottom rows are always the same shading; **position** – (a) the third and fourth rows move down one row, (b) the top row then takes on the shading of the bottom row

Test 5 (page 86)

1 e **rotation** – (a) the 'flag' rotates in a clockwise direction, (b) the circle rotates in an anticlockwise direction

2 d **proportion** – (a) the cross reduces in size moving left to right across the sequence, (b) the dashed square increases in size, moving left to right across the sequence; **rotation** – the triangle rotates in a clockwise direction around the square, moving left to right; **shading** – the triangle alternates between white and black

3 b **rotation** – (a) the white square with the arrow rotates clockwise around the square, moving left to right one-quarter of a side each time, (b) the black square with U-shape rotates clockwise around the square, moving left to right one-quarter of a side each time
Distractors: **shape** – the two central boxes do not change; **position** – the direction of the arrow and U shapes do not change in relation to their small squares

4 c **rotation** – (a) the curved line rotates clockwise by one side, moving left to right across the sequence, (b) the arrow rotates clockwise, from one corner to the next, moving left to right across the sequence; **shading** – the striped shading on the triangle alternates from side to side; **number** – of circles in the arrow tail decreases by one, moving left to right

5 d **proportion** – the shading on the circle increases by one-fifth each time (so the answer will have two-fifths of the circle shaded; **direction** – of the shading increases in a clockwise direction, moving around the circle
Distractors: **rotation** – the main teardrop shape rotates, although all the answer options are correct; **number** – of bars increases although all the answer options are correct

6 c **position** – (a) the triangle inverts in alternate pictures, (b) the part circle moves between the shaded vertices from striped corner to the white corner and then to the black corner and then repeats, (c) the diagonal shading alternates between parallel with the central triangle and parallel with the triangle base
Distractors: **rotation** – the elements all appear to rotate (as suggested by the answer options) rather than follow simple alternating patterns

7 a **number** – of lines around the corner square increases in every other step; **line style** – of the lines added alternates between solid and dashed; **direction** – of the diagonal shading points towards the corner of the square
Distractors: **rotation** – (a) the sequence involving the arrows rotating around the circle is a distraction, (b) the inner square rotates anticlockwise, moving left to right across the sequence (although all the answer options are correct)

8 b **rotation** – the black triangle rotates around the square in an anticlockwise direction; **line style** – of the large shape (in both top and bottom rows separately) follows the sequence dotted, dashed, solid; **number** – of small shapes increases by one (in both top and bottom rows separately); **shading** – the added shape alternates in colour

9 e **number** – of circles decreases by one in both the top and bottom rows separately; **shading** – of the triangle changes from black to white to diagonal stripes in sequence in both top and bottom rows; **position** – the small square moves down one-quarter of a side and then back up the same side again; **direction** – of the cross in the small box is different in the top and bottom rows
Distractors: **shading** – of the circles is white on the bottom row and black on the top row (although all the answer options are correct)

10 e **number** – (a) one side is added to the large hexagon across the sequence, working clockwise, (b) two sides are added to the small hexagon in the top row, working anticlockwise; **position** – (a) the black semi-circle in the top row moves round to fit on the last new side, (b) the white semi-circle in the bottom row matches the position of the black semi-circle in the next square, (c) the arrow points away from the white semi-circle

6 Matrices

Matrices using numbers and shapes (page 90)

1 The apple in the top right square should match the original picture but should be smaller or bigger. The fruit in the bottom row can be any fruit that is not an apple. This fruit changes size following the same rule as the apple. For example:

2 e moving left to right across the rows: **proportion** – (a) the left-hand shape enlarges, (b) the right-hand shape reduces; **line style** – of the two shapes swaps over

3 a **shape** – the large shape changes from one column to the next; **proportion** – the size of the triangle changes in proportion from one row to the next; **angle** – the line from the triangle to the outer box is horizontal in the first column but points 45° upwards in the second column

4 c **shape** – the shapes match in the diagonals from top left to bottom right; **proportion** – following the rows left to right, the number of shapes on either side of the central line changes in proportion: 3-1, 2-2, 1-3
Distractors: **position** – of the diagonal line in the boxes does not change

5 d **shape** – the top shape changes between rows; **number** – the number of times a line crosses the baton increases from left to right; **shading** – changes between rows; **proportion** – of the square shaded increases from left to right; **line style** – the thickness of the baton changes between rows
Distractors: **shape** – at the ends of the batons does not change

Matrices using position and direction (page 92)

1

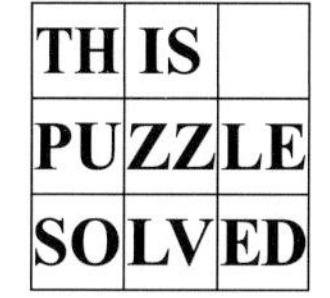

2 b **reflection** – the shapes reflect vertically through the centre of the grid; **position** – in the left-hand column, the bottom layer is the quadrilateral, the middle layer is the circle and the top layer is the L-shape but when reflected in the second column the middle layer moves to become the top layer

3 e **translation** – the shapes in the first two columns assemble in the final row but note that the shapes do not reflect, rotate or change shading: (a) the two shapes in the left-hand column sit one on top of the other on the left-hand side of the box, (b) the shape in the middle box is positioned on the right, (c) the lines from the first two boxes join the two sets of shapes together

4 c **rotation** – the arrow moves a quarter of the square anticlockwise, moving left to right and continuing onto the second and third rows as an extended sequence; **position** – the bands follow a pattern from top left to bottom right; **shading** – of the bands works from top left to bottom right

5 a **translation** – the shapes assemble along the diagonal, top left to bottom right with the complete shape in the central column; **position** – the orientation of the shape changes between the diagonals

Matrices combining different skills (page 94)

1 hexagon

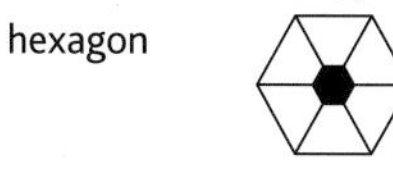

2 e **number** – the grid begins with four bold lines with one line being removed as the pattern works anticlockwise around the grid; **reflection** – the central shape reflects horizontally, moving left to right

3 c **number** – of small parallel lines around the edge increases in a clockwise direction; **shading** – (a) the shading of the circles is consistent working around the hexagons, (b) the shading of the squares alternates, working around the hexagons

4 d **position** – the shapes that appear in the second and third columns alternate between rows; **reflection** – the shapes in the middle two rows alternate and reflect (flip upside down); **shape** – the overlap of the two shapes creates the same type of shape as the smaller shape; **proportion** – (a) the large shape on the left reduces in size in the second square, (b) the large shape on the right reduces in size in the third square

5 b **translation** – the shapes in columns 2 and 3 combine to make the shapes in column 1 following the direction bottom right to top left; **proportion** – when the two shapes join together, the new shape (in column 1) is enlarged; **shading** – moves along the row with the shading from the bottom of the joined shape becoming the shading of the shape in column 2 and the shading from its top becoming the shading of the shape in column 3

Test 6 (page 96)

1 b **reflection** – the shapes reflect in a vertical line and move to the diagonal box; **shading** – of the shapes swaps

2 e **proportion** – the top right-hand shape in the box on the left reduces in size and moves into the left-hand shape (the shading does not change); **position** – the shapes at the top of the box move to bottom left; **rotation** – the shape at the bottom of the box rotates 135° anticlockwise and moves to the top of the box

3 c **position** – (a) the circles and crosses form an alternating pattern moving around the box, (b) the crosses touch the corners of the boxes, (c) the triangles and bullets also alternate in the outer edges of the box, (d) the bullets always have the straight edge pointing inwards and the triangles point inwards

4 b **translation** – the shapes translate from the top two rows in each column to the bottom row; **reflection** – the joined-together shapes in the bottom row reflect in a vertical line; **shading** – of these shapes swaps

5 a **reflection** – (a) the arrows reflect across the box, so the opposite corners reflect, (b) once reflected the arrows and circles reflect within the box (so the shading swaps sides); **line style** – matches along the diagonals from top left to bottom right; **shape** – the shape in the corner of the box matches the shape in the opposite corner of the central box

6 d **shape** – there is a triangle in each outer corner; **shading** – (a) the triangles alternate in colour with those in the boxes next to them, (b) an extra quarter is shaded in the small square, working in an anticlockwise direction, (c) the quarter shading is added in a clockwise direction

7 e **rotation** – the shapes within each small group of four rotate in a clockwise direction, moving around the outer shape in a clockwise direction
Distractors: **shape** – the shapes in the centre are the same as the shape in the opposite triangle (note that these shapes do not link with the squares around the edge, although at first sight they appear to do so)

8 a **number** – an extra circle is added, working clockwise around the hexagon; **shading** – as the new circle is added, the colour of the circles swaps (so the first is white/black, the next will be black/white and so on); **position** – of the shading in the long parallelogram alternates from top to bottom, moving around the hexagon; **proportion** – of the triangle shaded reduces by one-quarter moving around the hexagon

9 d **number** – of crosses increases working in an anticlockwise direction; **shape** – the shapes are identical in the same 'column'; **shading** – (a) the shading of the small shape is identical in the same 'column', (b) the shading of the left-hand shape is the same working diagonally from top right to bottom left

10 c **translation** – the top and middle rows join in the bottom row; **shading** – of the two shapes swaps

11+ sample test (page 98)

1 d **position** – all the other options have the triangle positioned with one vertex of the V in the Z-shape
In common/distractors: **rotation** – of the Z-shape is random, **reflection** – whether the Z-shape is reflected is random

2 e **position** – the dashed line in e is the only one that touches the other shapes
In common/distractors: **shape** – whether the shapes in each picture match or not is random; **rotation** – of the pictures is random; **line style** – whether the line is straight, bent or curved is random

3 a **number** – all the other options show lines that make two internal angles with each other, except for **a** where the three lines make four internal angles
In common/distractors: **angle** – the angles made by the lines are random; **proportion** – the length of the lines is random; **rotation** – of the pictures is random

4 c **number** – all the pictures have eight sides except for option **c**, which has ten
In common/distractors: **position** – (a) whether the shapes overlap is random, (b) the number of sides inside/outside the shape is random, (c) it is not important whether sides are parallel (or have symmetry); **shading** – random

5 c **position** – the thick band is always on the forward face of the patterned shape, working clockwise, except in **c**
In common/distractors: **number** – of circles is random; **position** – of the blank quarter is not important; **line style** – of the central circle (double or single) is random

6 c **shape** – (a) each picture has an outer square, (b) there is a small internal shape (random); **position** – the circle on the end of the line cuts across one corner next to the striped band; **number** – of sides in the small internal shape is the same as the number of lines on the side of the outer shape
In common/distractors: **shading** – on the side band does not change; **rotation** – whether the outer shape appears rotated is not important

7 d **shape** – (a) the small shapes are identical to each other, (b) the large shape is the same as the small shapes inside it; **number** – there are always three small shapes
In common/distractors: **shading** – of the shapes is not important

8 b **number** – there are always eight shaded boxes in the picture; **shape** – there is always a small square and small circle; **position** – (a) the square is always above the circle, (b) the boxes only connect on two edges; **shading** – runs from top right to bottom left

9 c **shape** – (a) the outer shapes have no straight edges; (b) there is always a circle with a cross and one smaller circle in each picture; **line style** – (a) there is always a double outer line, (b) the inner line is thin and solid

10 a **reflection** – each picture is made of a reflected outer shape; **shape** – (a) one shape contains a triangle, (b) the other shape contains a rectangle; **position** – the lower inside shape is always shaded black

11 a **proportion** – the top small shape enlarges; **position** – lower shape goes inside upper shape; **shading** – the new large shape takes its shading from the large shape on the left

12 e **reflection** – (a) the shape reflects horizontally, (b) the shape reflects vertically to create the final picture; **rotation** – the shading rotates by 90°

13 d **rotation** – the outer lines and corners reflect in a horizontal line; **proportion** – the bottom shape in the line of three small shapes reduces in size, the middle shape stays the same and the top shape increases in size; **position** – these three shapes fit one inside the other; **shading** – on the outer corner shape rotates 90° after it is reflected

14 e **proportion** – the shape on the left in the first picture reduces in size and moves beneath the shapes on the right in the second picture; **shape** – the concentric shapes in the second picture are the same as the outer shape in the first picture; **number** – the number of sides on the inner shape in the first picture equals the number of concentric shapes in the second picture; **shading** – (a) the shading in the shape on the left becomes the shading on the outer shape, (b) the shading from the central shape moves to the shape below

15 b **number** – the number of rules become the sides of the large shape; **rotation** – the small shape rotates 180° and moves inside the large shape; **shading** – the shading in the smaller shape moves to the larger shape; **position** – the black circle(s) move inside the smaller shape
In common/distractors: **number** – the shading lines look as if they could relate to the number of lines above, but they do not

16 d When the net is folded to make the cube the arrow and black face fold inwards to make the sides next to the white circle (using the idea of the blue arrows that make the elephant's head). Option **a** looks as if it could be correct but the black circle would be behind the cube when it is turned through 90° (this can be seen by the direction of the arrow).

17 a When the net is folded to make the cube the shapes on the central line will not change direction, so the heart must point to the blank face. The U-shape folds inwards (using the idea of the red arrow that makes the elephant's cheeks). There are no other options where the direction of the shapes in the nets gives a correct answer.

18 b When the net is folded to make the cube the shapes on the central line wrap around to meet and so are all pointing in the same direction. Therefore the two U-shapes are the same way up. The row of three circles folds inwards (using the idea of the green arrow that makes the elephant's middle) to touch the bottom of one U-shape. There are no other options where the direction of the shapes in the nets gives a correct answer.

19 c The final shape has a square base, so there will need to be four triangles, forming the top pyramid, and five squares to make the sides and the base. The triangles need to come to a point in the centre for the pyramid to work (they do not do this in option **e**). Option **c** is the only possible answer when considering the triangles.

20 c Looking at the left-hand stack, it has two cubes at the front, then one on the right on the third layer. This means the left-hand side of the 2D plan must have two squares on the left and one square on the right. This can be seen as the upside-down L-shape in options **a** to **d**. The right-hand stack has three cubes on the base, three cubes going up and one cube sticking out. So, on the plan you will see three cubes along and two cubes going up on the right-hand side as shown in option **c**.

21 e The stack of cubes has two cubes at the front with a gap between them (they are also at two different levels) so the plan will have a square at the top and the bottom of the first column. Behind these cubes are two cubes on the same level (which will be the two bottom squares in column 2 of the plan). There is a further cube on the base behind these cubes with the two final cubes on the next level. This will give a plan with two squares at the bottom of column 3 and one square in the centre of column 4.

22 b The fold lines act like a diagonal reflection so, when the square is unfolded, the circles will be parallel with the vertical side of the square, with the triangles sitting alongside each other with their flat bases facing downwards.

23 e The fold lines act like line lines of reflection so, when unfolding the second fold that was made, the triangle and circle will reflect horizontally and therefore be upside down, with the triangle at the top left of the square. Unfolding the first fold that was made along the vertical line will mirror the pattern.

24 a The fold lines act like line lines of reflection so, when unfolding the third fold that was made the heart will reflect vertically in the centre of the 'triangle'. This means that the second heart will point to the left (the circle is not punched through the triangle). When the second (horizontal) fold that was made is unfolded, there will be a second circle above the first. Now, unfolding the diagonal fold, the pattern of two circles and two hearts will be reflected in this line.

25 d The fold lines act like line lines of reflection so, when unfolding the third fold that was made the circles will reflect in a horizontal line to create two more circles above. When the second (diagonal) fold is opened up, a second heart will appear pointing to the right and the four circles will move to the bottom right. When the final horizontal fold is opened, this whole pattern will reflect in the horizontal line.

26 b **rotation** – the picture is rotated 135°; **position** – (a) rectangle points away from the domed shape, (b) the cross touches the sides rather than the corners of the domed shape; **shading** – the section of the rectangle outside the main shape is shaded

27 d **rotation** – the picture is rotated 135°; **position** – (a) the cross touches the corners of the small square, (b) the black triangle overlaps the corner of the large shape and is the top layer, (c) the position of the cross containing the square is on the corner to the right of the black arrow; **shading** – on the central triangle is at right-angles to its base
Distractors: **shape** – the black lines and white semi-circles are not important

28 e **rotation** – the picture is rotated 180°; **position** – (a) the double arc points towards the curve of the '3'-shape, (b) the circle is at the end of the straight edge of the '3'-shape while the diamond is at the bottom; **shading** – the circle is white and the diamond is black; **proportion** – the second straight line on the '3'-shape is approximately two-thirds the length of the first

29 c **reflection** – the complete picture is reflected in a vertical mirror line so instead of pointing to the top left the new picture points to the right
Distractors: **position** – (a) the cross in the square touches the corners, (b) the position of the black horizontal band is in the middle of the tail of the '7'-shape; **shading** – the bottom corner of the triangle is shaded

30 e **reflection** – the complete picture is reflected in a vertical mirror line so instead of pointing to the top left the new picture points to the top right
Distractors: **reflection** – the band on the long edge of the shield should not reflect horizontally; **position** – the white circle and square should be placed against the bands at the top edge

31 d **reflection** – the complete picture is reflected in a vertical mirror line so the diagonal shading runs in the opposite direction, the triangle points to the right and the circle is on the left
Distractors: **position** – the black circle should always be behind the main shape; **shape** – (a) the white rectangle does not change shape, (b) there is always a gap between the white rectangle and the shape above

32 d Shape **d** is rotated 90° anticlockwise around a vertical line and then rotated 180° around a horizontal line (flipped over).

33 e Shape **e** is tipped over to the left and then rotated 90° anticlockwise.

34 a Shape **a** is rotated 90° clockwise around a vertical line.

35 c Shape **c** is tipped up to stand on the two-block cube that is currently pointing to the right and then rotated 90° anticlockwise around a vertical line.

36 b Shape **b** is flipped over by 180°.

37 e 45° is half a right-angle. Following the point of the triangle (as if it is the hand on a clock) gives e as the correct answer. The angle of the shading also rotates in the same way.

38 c 135° is one and a half right-angles. Rotating the picture by 90° would give option **d**. Following the position of the four squares makes **c** correct as it is clearly rotated more than 90° and less than 180°.

39 b 315° is 45° less than a complete 360° turn. Therefore the answer will be a 45° anticlockwise rotation. The change in position of the line at the base of the shape is an easy feature to follow in this rotation.

40 e The angle at the bottom of the shape is one of the most identifiable features. This fits into the angle at the bottom left of option **e**. The shape extends to the second vertical line within the large picture.

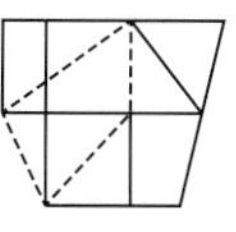

41 d The most identifiable feature of the small shape is the long angled side on the right. Option **d** is the only shape with this angled line.

42 e The most identifiable feature of the small shape is the triangle section at the bottom. Options **a**, **b** and **c** do not include this shape. Option **d** does not include the angled line at the top of the shape, so this leaves option **e**.

43 e The small shape is made up of two quadrilaterals. Taking the left-hand quadrilateral, options **c** and **d** can be rejected because neither has a shape with a narrow horizontal edge that would match the top of this shape. In **a** and **d** there is no right-angle that matches the right-hand side of the shape, so the answer must be option **e**.

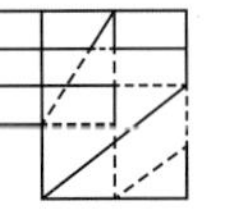

44 a The T-shaped block has been rotated around a vertical line by 90° anticlockwise and then rolled forward by another 90° to sit on top of the two-cube block (which has fallen over to lie flat). The L-shaped block is rotated 90° around a vertical line and then dropped backwards 90° to fit against the side of the two-cube block.

45 d The two long L-shaped blocks have been joined together to form a long rectangle at the back. The short L-shaped block has been rotated 90° around a vertical line and then rolled backwards 90° to sit against this rectangle. The single block then sits alongside on the right.

46 e The T-shaped block does not change position and a single cube is placed beside it on the right. The other single cube is placed on the right-hand bar of the T. The L-shaped block rotates 90° clockwise around a vertical line and then rolls backward 90° around a horizontal line to sit against the back of the T. The three-cube block then stands on its end to the left of the L-shape.

47 b **line style** – the first letter represents the line style of the outer shape: L is a dashed line and M a solid line; **shape** – (a) the second letter represents the outer shape: R a triangle, S a circle and T a square, (b) the third letter represents the inner shape: X is a circle and Y a triangle

48 b **shape** – (a) the first letter represents whether the outer shape is open or closed: R is open, S is closed, (b) the second letter represents the small shape: X is a square, Y a triangle and Z a quadrilateral
Distractors: **number** – of small circles is not important

49 e **shading** – the first letter represents the shading style in the rectangle: L diagonal (top right to bottom left), M horizontal and N diagonal (top left to bottom right); **line style** – the second letter represents the line style of the outer box: X long dash and Y short dash; **position** – the third letter represents the position of the black circle: P at the top of the square and Q at the bottom
Distractors: **position** – of the white circle is not important

50 c **shape** – the first letter represents the shape of the large triangle: F is a right-angled triangle and G an isosceles triangle; **shading** – the second letter represents the shading at the top of the triangle: P is diagonal and Q is black; **shape** – the third letter represents the inner shape: W is a circle , Y a triangle and Z a square
Distractors: **number** – of black circles is not important; **position** – of the small triangles is not important

51 d **number** – the first letter represents the number of small parallel lines: X is two, Y is three and Z is four; **shape** – the second letter represents the shape created by the longer lines: F is an L-shape, G a single line and H a U-shape; **line style** – the third letter represents the line style of the shape made by the longer lines: L is a double line, M a triple line and N a single line
Distractors: **shape** – the triangles are not relevant to the answer; **position** – of all the elements is not important

52 b **shape** – the first letter represents the larger shape inside the outer circle: R square, S hexagon and T triangle; **line style** – the second letter represents the line style of the outer circle: W is thin and solid, X long dash, Y short dash and Z thick and solid; **proportion** – the third letter represents the proportion of the small circle that is shaded: L half, M quarter and N three-quarters
Distractors: **position** – the rotation/position of the large shape is not important; **shape** – the small triangles and small circles are not important

53 c **rotation/number** – (a) the number of sides of the outer square reduces by one in a clockwise direction, (b) the number of sides of the inner square reduces by one in a clockwise direction; **position** – the black triangle alternates between the bottom and top of the small square

54 d **rotation** – (a) the semi-circle rotates one-quarter turn anticlockwise working left to right across the sequence, (b) the segment rotates one side of the square clockwise, working left to right across the sequence; **line style** – the semi-circle alternates between a long dashed line and a short dashed line; **shading** – the colour of the square and segment alternates

55 a **position** – (a) the black circle alternates between being inside and outside the hexagon, (b) the end of the bar moves alternately between the top right and bottom left corners of the square; **number** – of V-shapes around the vertices increases by one, moving left to right across the sequence; **rotation** – (a) the striped triangle moves anticlockwise around the shape, moving left to right across the sequence, (b) the black triangle moves clockwise around the shape (so the black and striped triangles are, in fact, on top of each other in the answer option and therefore the striped triangle is not visible in the answer)
Distractors: **size** – of the hexagons alternates from large to small, although all the answer options are the correct size

56 e **rotation** – (a) the triangle moves 90° clockwise, working left to right across the sequence, (b) the open circle moves 90° clockwise, (c) the square moves 90° anticlockwise, (d) the circle with a cross moves 90° anticlockwise; **shading** – the central circle alternates between black and white

57 b **shape** – the triangle and rectangle alternate between squares; **rotation** – (a) the whole picture rotates 45° anticlockwise, working left to right across the sequence, (b) the triangle flips upside down when it appears (so it points first away and then towards the central line; **position** – the shading swaps sides on the semi-circle from one square to the next; **line style** – follows a sequence of three: thick solid line, thin solid line, dotted line

58 e **number** – of small bands increases by one, moving left to right as well as up and down the rows; **position** – (a) the small bands begin on the left-hand edge on the base of the shape (making option **a** incorrect), (b) the arrow points away from the triangle; **rotation** – (a) the shape on the top row rotates one-quarter turn clockwise, (b) the shape on the bottom row rotates one-quarter turn anticlockwise

59 d **rotation** – the whole picture rotates 180°; **position** – the small square moves to the opposite side of the picture; **shading** – the medium- and large-sized shapes swap shading

60 d **rotation** – the outer shape rotates one-quarter turn clockwise, moving left to right across the rows (as can be seen by the movement of the dashed lines); **line style** – changes between rows: short dash, long dash, then dotted; **shading** – follows a pattern running top right to bottom left; **shape** – the shapes follow a pattern top left to bottom right

61 a **shape** – the shapes work in a diagonal pattern top right to bottom left (the two rotations of the lozenge are counted as different shapes); **number** – of small white circles increases working down the rows; **position** – of the small circles is in the same pattern on each row, **shading** – works in a pattern top left to bottom right (because the shapes rotate, the shading appears to change but is, in fact, the same)

62 e **shape** – the kind of shape changes between rows; **number** – the pattern works top right to bottom left (for example, a four-sided shape in the top row, four squares in the middle row, a four-pointed star in the bottom row); **shading** – also works in a pattern from top right to bottom left: the first diagonal has one-quarter shaded, the second has half shaded and the third has three-quarters shaded. In the middle row the proportions relate to the number of shapes, so half the shapes should be shaded; **position** – the position of the shaded shapes/sections of shapes follows the diagonal top right to bottom left so, because the top half of the central hexagon is shaded in the top row, the top three hexagons will be shaded in the central row

63 d **position** – there are always two triangles positioned inside the large triangle, moving around the shape; **shading** – (a) the small outer triangle is always black, (b) the inner triangle alternates between black and white, (c) the two inner concentric circles alternate in colour between black and white; **rotation** – the small black section in the outer circle moves one-sixth of a turn clockwise

64 b **number** – of sides in the main shape increases by one in each row from the base (three, four, five, six), excluding the central shape; **shading/pattern** – picked up from the opposite side of the central shape